Lecture Notes in Mathematics

Edited by A. Dold and B. Eckmann

1373

Georgia Benkart J. Marshall Osborn (Eds.)

Lie Algebras, Madison 1987

Proceedings of a Workshop held in Madison, Wisconsin, August 23–28, 1987

Springer-Verlag

Berlin Heidelberg New York London Paris Tokyo Hong Kong

Editors

Georgia Benkart
J. Marshall Osborn
Department of Mathematics, University of Wisconsin
Madison, WI 53706-1388, USA

Mathematics Subject Classification (1980): Primary: 17B20, 17B50, 17B67
Secondary: 16A61, 17B56, 17B65, 18G15

ISBN 3-540-51147-4 Springer-Verlag Berlin Heidelberg New York
ISBN 0-387-51147-4 Springer-Verlag New York Berlin Heidelberg

Printing and binding: Druckhaus Beltz, Hemsbach/Bergstr.
2146/3140-543210

Preface

During the academic year 1987-1988 the University of Wisconsin, Madison hosted a Special Year of Lie Algebras. A Workshop on Lie Algebras, of which these are the proceedings, inaugurated the special year. The principal focus of the year and of the workshop was the long-standing problem of classifying the simple finite dimensional Lie algebras over algebraically closed fields of prime characteristic. However, other lectures at the workshop dealt with the related areas of algebraic groups, representation theory, and Kac-Moody Lie algebras.

The titles of the fourteen papers presented at the workshop can be found at the end, followed by a list of the participants in the workshop and their addresses. Nine of these papers, (eight research articles and one expository article), comprise this volume. The first paper, by Strade, develops the notion of the absolute toral rank of a modular Lie algebra. This new concept combines earlier approaches of Block-Wilson and Benkart-Osborn, and it seems to play a critical role in determining the structure of simple Lie algebras of prime characteristic. The next three papers investigate various topics related to the classification problem: embeddings of generalized Witt algebras; isomorphism classes of Hamiltonian Lie algebras; and Lie algebras with subalgebras of codimension one and their relationship to forms of Zassenhaus Lie algebras. The determination of the restricted simple Lie algebras over algebraically closed fields of prime characteristic has been accomplished recently by Block and Wilson. Serconek and Wilson use this result as the starting point for their discussion of forms of restricted simple algebras. In the next paper Varea employs the classification of the rank one simple Lie algebras to investigate the subalgebra lattice of supersolvable Lie algebras. The final three papers treat problems in Kac-Moody algebras.

We would like to take this opportunity to express our deep appreciation to the National Science Foundation for its support (through grant #DMS 87-02928) of the Special Year of Lie Algebras. Without its support this workshop and the other activities of the special year would not have been possible. Thanks also go to Rolf Farnsteiner, David Finston, Thomas Gregory, Helmut Strade, and Robert Wilson for their participation in the events of the special year, and to the typists, Dee Frana and Diane Reppert, and referees who helped in the preparation of this volume.

Georgia Benkart and J. Marshall Osborn June 1, 1988

Table of Contents

Table of Contents

THE ABSOLUTE TORAL RANK OF A LIE ALGEBRA

Helmut Strade

Abstract: The new concept of an absolute toral rank for subalgebras in arbitrary modular Lie algebras is introduced. All nonsimple Lie algebras of absolute toral rank ≤ 2 are determined in terms of smaller constituents. The final result is the first step towards the classification of all simple modular Lie algebras.

Introduction.

In the theory of modular Lie algebras there are several fundamental concepts and objects based on the notion of a "toral rank". In a restricted Lie algebra, for example, the toral rank of a torus gives the GF(p)–dimension of the root lattice determined by this torus. Tori of maximal toral rank occur in the classification theory as very important objects. In the theory of non–restricted semisimple Lie algebras the toral rank of a Cartan subalgebra (CSA) is a frequently used concept, and the particular case that a Lie algebra has a CSA of toral rank one is a very central one.

In this note I shall introduce in §1 the concept of an "absolute toral rank TR(G,L) of a subalgebra G in the Lie algebra L" for arbitrary Lie algebras over any field of characteristic $p \neq 0$. This generalizes all of the above–mentioned concepts to arbitrary Lie algebras. Important tools in this context are the concepts of a "restrictable Lie algebra" and a "p–envelope" of an arbitrary Lie algebra [5] as well as their structure theory [7].

In §2 we will prove several results on the absolute toral rank, which generalize well–known results on the dimension of tori in restricted Lie algebras.

In §3 we consider subalgebras $C_L(T)$ of a modular Lie algebra L for a torus T in some p–envelope of L. The main result in this section is a far–reaching extension of [8, Theorem 2.1] which yields that under suitable conditions subalgebras of this type act triangulably on L.

Partially supported by NSF grant No. DMS–8702928.

Applications are given in section 4. There we describe all Lie algebras of absolute toral rank ≤ 2 in terms of: solvable algebras, algebras having toral rank 1 with respect to some CSA, and simple algebras of absolute toral rank 2 satisfying some additional assumption concerning their CSA's. The final Theorem 4.8 may be considered a generalization of [3, Theorem 4.1.1] to non-restricted algebras. All Lie algebras under consideration are finite dimensional over an algebraically closed field F of characteristic $p > 0$.

This note extends results of a talk, which was given on the occasion of the opening workshop of the "Special Year on Lie Algebras" in August, 1987 at Madison, Wisconsin.

§1. The toral structure of a Lie algebra.

In this chapter we will use the concept of a p-envelope [5] in order to transfer some methods, which are very fruitful in the case of restricted Lie algebras, to arbitrary algebras.

For convenience we use in this note the following abbreviation:

Let L be a restricted algebra. Put

$$MT(L) := \max\{\dim T \mid T \text{ is a torus of } L\} .$$

Using the notation $C_X(Y) = \{x \in X \mid [Y,x] = 0\}$ we recall

Lemma 1.1: Let L be a restricted Lie algebra. Suppose that T is a torus of L.

1) Any T-invariant subspace $W \subset L$ decomposes

$$W = C_W(T) + [T,W] .$$

2) If I is a restricted ideal of L and L/I is a torus, then there exists a torus $T' \supset T$ such that $L = T' + I$.

Proof: [7,(II.4.4),(II.4.5)] □

Lemma 1.2: Let L be restricted.

1) If I is a restricted ideal of L, then

$$MT(L) = MT(L/I) + MT(I) .$$

2) Suppose that K_1 is a restricted subalgebra and K_2 is a restricted ideal of L. Then

$$MT(K_1+K_2) = MT(K_1) + MT(K_2) - MT(K_1 \cap K_2) .$$

Proof: 1) Let $\pi: L \to L/I$ denote the canonical homomorphism.

For any torus T of L, $\pi(T)$ and $T \cap I$ are tori of L/I and I, respectively. This shows that $MT(L) \leq MT(L/I) + MT(I)$. In order to obtain the reverse inequality we apply Lemma 1.1: Let T be a torus of I and R a torus of L/I. Then L has a torus T' containing T with $T' + I = \pi^{-1}(R)$ (put in Lemma 1.1 (2) $\pi^{-1}(R)$ for L). Therefore

$$\dim(T') \geq \dim \pi(T') + \dim T' \cap I \geq \dim R + \dim T .$$

This gives the result.

2) Applying 1) we obtain

$$MT(K_1+K_2) = MT(K_1+K_2/K_2) + MT(K_2)$$

$$= MT(K_1/K_1 \cap K_2) + MT(K_2) = MT(K_1) - MT(K_1 \cap K_2) + MT(K_2). \square$$

A p-envelope of an arbitrary modular Lie algebra L is a triple $(H,[p],i)$ consisting of a restricted Lie algebra $(H,[p])$ and an embedding $i: L \to H$ such that $i(L)$ generates H as a restricted subalgebra [5]. We often consider L a subalgebra of H and call H a p–envelope of L. The following notation will be used: L is always a finite dimensional Lie algebra and L_p denotes a finite dimensional p–envelope containing L as a subalgebra. If L and L_p are given and G is a subalgebra of L, then G_p denotes the minimal restricted subalgebra of L_p which contains G.

Remarks:

1) Suppose that L is nilpotent. Then L_p is nilpotent. L_p has a unique maximal torus. This torus is contained in $C(L_p)$. It is the set of all semisimple elements [7,(II.1.3), (II.4.2)].

2) Every representation $\rho: L \to gl(V)$ can be extended to a representation $\hat{\rho}: L_p \to gl(V)$ [7,(V.1.1)].

3) Suppose that G is a subalgebra of L and $V \subset L$ is G–invariant. Then V is G_p–invariant. (This is a direct consequence of 2).)

4) Suppose that $\hat{\rho}$ is a restricted representation, such that $\hat{\rho}(t)$ is a nilpotent transformation for every semisimple element t. Then $\hat{\rho}(L_p)$ consists of nilpotent transformations.

Proof of 4): For any $x \in L_p$ there exists a power $t := x^{[p]^r}$ such that t is a semisimple element of $(L_p,[p])$. (We call this the semisimple part of the element x [7,(II.3.5)]). Our present assumption implies that $\hat{\rho}(t)$ is nilpotent and $\hat{\rho}(x)^{p^r} = \hat{\rho}(x^{[p]^r}) = \hat{\rho}(t)$. Then $\hat{\rho}(x)$ is nilpotent. □

It is well–known that for a semisimple element t, $\hat{\rho}(t)$ is nilpotent if and only if $\hat{\rho}(t) = 0$.

Some main results on p–envelopes are summarized as follows.

Theorem 1.3 ([7]): Let L be a finite dimensional Lie algebra.

1) L possesses a finite dimensional p–envelope. If L is semisimple, it has a finite dimensional semisimple p–envelope.

2) Any two p–envelopes of L of minimal dimension are isomorphic as ordinary Lie algebras.

3) Let $(H_k,[p]_k,i_k)$ (k=1,2) be p–envelopes of L. Then there exists a (non–restricted) homomorphism $f\colon H_1 \to H_2$ and a subspace $J \subset C(H_2)$ such that

$$H_2 = f(H_1) \oplus J, \quad f \circ i_1 = i_2, \quad \ker(f) \subset C(H_1).$$

Proof: 1) [7,(II.5.6)], 2) [7,(II.5.8)] 3) [7,(II.5.6), (II.5.7), (II.5.5)]. □

Corollary 1.4([4]): Let $(H_k,[p]_k,i_k)$ (k=1,2) be two p–envelopes of L. Then there exists a restricted isomorphism

$$\varphi\colon H_1/C(H_1) \xrightarrow{\sim} H_2/C(H_2)$$

with $\varphi(i_1(x) + C(H_1)) = i_2(x) + C(H_2)\ \forall x \in L$.

Proof: We employ the notation and the result of Theorem 1.3 (3). Let x be an element of H_1 with $f(x) \in C(H_2)$. Then $f([x,i_1(L)]) = 0$ and hence $[x,i_1(L)] \subset i_1(L) \cap \ker(f) \subset \ker(i_2) = (0)$. This implies, as H_1 is generated by $i_1(L)$, $x \in C(H_1)$. Therefore f induces an isomorphism

$\varphi: H_1/C(H_1) \longrightarrow H_2/C(H_2)$ with $\varphi(i_1(x) + C(H_1)) = i_2(x) + C(H_2)\ \forall x \in L$.

Moreover for any $h_2 \in H_2$ there exist $h_1 \in H_1$ and $z \in C(H_2)$ such that $h_2 = f(h_1) + z$. Hence we obtain for any $x \in H_1$

$$[f(x^{[p]_1}) - f(x)^{[p]_2}, h_2] = [f(x^{[p]_1}), f(h_1)] - (\text{ad } f(x))^p(f(h_1)) = 0.$$

This shows that $f(x^{[p]_1}) - f(x)^{[p]_2} \in C(H_2)$. Then φ is a restricted isomorphism. □

The above result ensures, that the ensuing definition does not depend on the choice of the p–envelope.

Definition: Let L be a Lie algebra and $(H,[p],i)$ a p–envelope of L. Suppose that G is a subalgebra of L and G_p is the restricted subalgebra of H generated by $i(G)$.

1) $TR(G,L) := \max\{\dim T \mid T$ is a torus of $G_p + C(H)/C(H)\}$ is called the (absolute) toral rank of G in L.

2) For $G = L$ we call $TR(L) := TR(L,L)$ the absolute toral rank of L.

Remarks 1.5:

1) The above definition of the absolute toral rank of a Lie algebra L coincides with that given in [4].

2) If L is restricted and G is a torus of L then $TR(G,L) = \dim G/G \cap C(L)$ is the well–known toral rank of G in L. If L is restricted then $TR(L) = MT(L) - MT(C(L))$, so if L is centerless then $TR(L) = MT(L)$.

3) Suppose that G is a nilpotent subalgebra (or even a CSA of L). Let H denote a p–envelope of L and G_p the restricted subalgebra of H generated by G. Since G_p is nilpotent it has a unique maximal torus T. Then $\dim T/C_T(L)$ is called the toral rank of L with respect to G ([8]). Note that, since L generates H as a restricted algebra,

$$C_T(L) = C_T(H) = T \cap C(H)\,.$$

Therefore,

$$\dim T/C_T(L) = TR(T,H) = TR(G,L)$$

in this case.

In [8] Wilson gave a further description of the toral rank of L with respect to G in case that

G is a CSA of L. Then

$$L = \sum_{\alpha \in \Delta} L_\alpha(G)$$

decomposes into root spaces with respect to G. The dimension of the GF(p)–vector space spanned by Δ is the toral rank of L with respect to G. The same holds if G is nilpotent but not necessarily a CSA of L. This result yields a useful interpretation of TR(G,L) if G is nilpotent.

The concept of an "absolute toral rank of G in L" is the adequate concept which generalizes and unifies several definitions involving the term of a "toral rank". G. B. Seligman recently asked for a clear distinction between the various notions of "toral rank" and "rank". Our definition is a reply to this request.

§2. Properties of the absolute toral rank.

In the following we will prove some properties of the absolute toral rank. Note that for restricted semisimple Lie algebras TR and MT coincide. The following results are generalizations from this more specific situation.

Proposition 2.1.: Let L be a Lie algebra, G a subalgebra, L_p a p–envelope and G_p the p–subalgebra of L_p generated by G.

1) $TR(G,L) = TR(G,L_p) = TR(G_p,L_p)$, so that $TR(L) = TR(L_p)$.
2) $TR(G,L) = MT(G_p/C(L_p) \cap G_p) = MT(G_p) - MT(C(L_p) \cap G_p)$.

Proof: 1) follows directly from the definition.
2) The first equation follows directly from the definition of TR and MT. The second equation is a consequence of Lemma 1.2(1). □

Proposition 2.2: Let $G \subset K \subset L$ be Lie algebras.

1) $TR(G,K) \leq TR(G,L)$
2) $TR(G,L) \leq TR(K,L)$
3) $TR(G) \leq TR(K)$.

Proof: Let L_p denote a p-envelope of L and $G_p \subset K_p \subset L_p$ the restricted subalgebras generated by G and K, respectively. Apply Proposition 2.1(2).

1) The inclusion $C(K_p) \cap G_p \supset C(L_p) \cap G_p$ yields 1).

2) Let $T \subset G_p$ be a torus such that $\dim T/T \cap C(L_p)$ is maximal. T is also a torus of K_p. By definition we obtain $TR(K,L) \geq \dim T/T \cap C(L_p) = TR(G,L)$.

3) Putting $K = G$ and $L = K$ we obtain from 1), 2) $TR(G) \leq TR(G,K)$ and $TR(G,K) \leq TR(K)$, respectively.

□

The following is a refinement of [4,(1.2)].

Proposition 2.3: Let I be an ideal of K and K a subalgebra of L. Suppose that L_p is a p-envelope of L and let K_p denote the p-envelope of K in L_p. Put $J := \{x \in K_p \mid [x,K] \subset I\}$. Then

$$TR(K,L) = TR(K/I) + TR(J,L_p) \geq TR(K/I) + TR(I,L) \geq TR(K/I) + TR(I) .$$

Proof: a) We consider K contained in K_p. Then I is an ideal of K_p and J is a p-ideal of K_p. Observe that there is a canonical isomorphism

$$\varphi: K_p/I \,/\, C(K_p/I) \xrightarrow{\sim} K_p/J .$$

Since K_p is restricted, the homomorphic image K_p/I is restrictable [7,(II.2.4)]. Choose any p-mapping $[p]'$ on K_p/I and let G be the restricted subalgebra generated by K/I and $[p]'$.

For any $x \in K_p$ the element

$$(x+I)^{[p]'} - (x^{[p]}+I)$$

centralizes K_p/I. Thus we proved the implication

$$x + I \in G + C(K_p/I) \Rightarrow x^{[p]} + I \in G + C(K_p/I) .$$

As $K/I \subset G$, this shows that

$$K_p/I = G + C(K_p/I)$$

and hence

$$G \cap C(K_p/I) = C(G) .$$

Therefore

$$G/C(G) \cong K_p/I \,/\, C(K_p/I) \,.$$

The above remark also proves that an element $(x+I)^{[p]'} + C(G)$ is mapped under this isomorphism onto $(x^{[p]}+I) + C(K_p/I)$. Applying φ we obtain a restricted isomorphism

$$G/C(G) \xrightarrow{\sim} K_p/J \,.$$

b) Since $C(L_p) \cap K_p = C(L_p) \cap J$ and G is a p-envelope of K/I, application of Proposition 2.1 and Lemma 1.2 yields

$$TR(K,L) = MT(K_p) - MT(C(L_p) \cap K_p) = MT(K_p/J) + MT(J) - MT(C(L_p) \cap J)$$

$$= MT(G/C(G)) + TR(J,L_p) = TR(K/I) + TR(J,L_p) \,.$$

The remaining inequalities follow from the fact that $I \subset J$ and Proposition 2.2. □

<u>Proposition 2.4</u>: Suppose that I is an ideal of L. Then $TR(I,L) = TR(I)$.

<u>Proof</u>: Let L_p be a p-envelope of L and I_p the p-subalgebra generated by I. Take any torus T of I_p. Then L_p decomposes

$$L_p = C_{L_p}(T) + [T,L_p] \,.$$

Since I_p is an ideal of L_p, we have $[T,L_p] \subset I_p$. Thus

$$T \cap C(I_p) \subset T \cap C(L_p) \subset T \cap C(I_p) \,.$$

We conclude that $TR(T,L_p) = TR(T,I_p)$. The definition now yields $TR(I,L) = TR(I)$. □

<u>Proposition 2.5</u>:

1) Suppose that S_1, S_2 are Lie algebras. Then

$$TR(S_1 \oplus S_2) = TR(S_1) + TR(S_2)$$

2) Suppose that S_1, S_2 are ideals of a Lie algebra L. Then

$$TR(S_1+S_2) + TR(S_1 \cap S_2) \le TR(S_1) + TR(S_2)$$

3) Suppose that S_1 is a subalgebra and S_2 is an ideal of L. Then

$$TR(S_1+S_2) + TR(S_1 \cap S_2) \le TR(S_1, S_1+S_2) + TR(S_2) .$$

Proof: We prove 3) first. Let L_p be a p–envelope of L and K_i the p–algebra generated by S_i (i=1,2). Then $K := K_1 + K_2$ is a restricted subalgebra of L_p, generated by $S_1 + S_2$. Hence K is a p–envelope of $S_1 + S_2$.

According to Lemma 1.2 we have

$$MT(K) = MT(K_1) + MT(K_2) - MT(K_1 \cap K_2) .$$

Application of Proposition 2.1(2) to this equation yields

$$TR(K) + MT(C(K)) = TR(K_1,K) + MT(C(K) \cap K_1)$$
$$+ TR(K_2) + MT(C(K_2)) - TR(K_1 \cap K_2) - MT(C(K_1 \cap K_2)) .$$

Since $TR(K_1 \cap K_2) \ge TR(S_1 \cap S_2)$ we obtain from Proposition 2.1(1) and this equation

$$TR(S_1,S_1+S_2) + TR(S_2) - TR(S_1+S_2) - TR(S_1 \cap S_2)$$

(*) $$\ge TR(K_1,K) + TR(K_2) - TR(K) - TR(K_1 \cap K_2)$$

$$= MT(C(K)) - MT(C(K) \cap K_1) - MT(C(K_2) + MT(C(K_1 \cap K_2)) .$$

Let T_1,T_2 denote the unique maximal tori of $C(K) \cap K_1$, $C(K_2)$, respectively. Note that, as K_2 is an ideal of K,

$$[T_2,K] = [T_2^{[p]},K] \subset (\mathrm{ad}\, T_2)^p(K) \subset [T_2,K_2] = 0 .$$

Then $T_2 \subset C(K)$ and $T_1 + T_2$ is a torus of $C(K)$. Since $T_1 \cap T_2$ is a torus of $C(K_1 \cap K_2)$, we have

$$MT(C(K)) + MT(C(K_1 \cap K_2)) \ge \dim (T_1+T_2) + \dim(T_1 \cap T_2) = \dim T_1 + \dim T_2$$
$$= MT(C(K) \cap K_1) + MT(C(K_2)) .$$

Thus the right hand side of (*) is nonnegative. This proves 3).

2) is a direct consequence of 3) and Proposition 2.4, putting $I = S_1$.

1) Take in Proposition 2.3 $S_1 + S_2$ for K and L and S_1 for I. Observe that $S_1 + S_2/S_1 \cong S_2$. Then Proposition 2.3 yields

$$TR(S_1 \oplus S_2) \geq TR(S_2) + TR(S_1) .$$

In combination with 2) we obtain 1). □

Some subalgebras are of major importance in the classification theory.

Definition: a) Let L be a Lie algebra, L_p a p–envelope of L and T a torus of L_p. Decompose L into eigenspaces with respect to T

$$L = \sum_{\alpha \in \Delta} L_\alpha(T) .$$

A subalgebra

$$K = \sum_{\alpha \in \phi} L_\alpha(T), \quad \Phi \subset \Delta \subset T^*$$

is said to be a k–section (with respect to T) if ϕ is the GF(p)–vector space which is spanned by GF(p)–independent roots $\alpha_1, \cdots, \alpha_k$ relative to T.

b) Let H be a nilpotent subalgebra of L, H_p the p–subalgebra of L_p generated by H and T_0 the unique maximal torus of H_p. A subalgebra K is called a k–section with respect to H, if it is a k–section with respect to T_0.

Theorem 2.6: Let L be a Lie algebra and L_p a p–envelope of L. Suppose that $T \subset L_p$ is a torus with $TR(T,L_p) = TR(L)$. If K is a k–section with respect to T then

$$TR(K) \leq k .$$

Proof: Let R be the maximal torus of $C(L_p)$. Then $T + R$ is a torus with $TR(T+R,L_p) = TR(L)$ and K is a k–section with respect to $T + R$. Hence we may assume that T contains the maximal torus of $C(L_p)$. Let $K_p \subset L_p$ be the p–subalgebra generated by K and T_0 a torus of K_p. We have to prove that $\dim T_0/C(K_p) \cap T_0 \leq k$. Put $T_1 := C_T(K)$. Since $[T_1,K] = 0$ we have $[T_1,K_p] = 0$ and $[T_1,T_0] = 0$. Therefore $T_0 + T_1$ is a torus of L_p, proving

$$\dim (T_0 + T_1)/C(L_p) \cap (T_0+T_1) \le TR(L) .$$

Observe that $C(L_p) \cap (T_0+T_1)$ is a torus in $C(L_p)$ and hence is contained T. Then

$$C(L_p) \cap (T_0+T_1) \subset C(L_p) \cap T .$$

Thus we obtain

$$(*) \quad \begin{aligned} \dim T - \dim C(L_p) \cap T = TR(L) &\ge \dim(T_0+T_1) - \dim C(L_p) \cap (T_0+T_1) \\ &\ge \dim T_0 + \dim T_1 - \dim(T_0 \cap T_1) - \dim C(L_p) \cap T . \end{aligned}$$

Let $\alpha_1,\dots,\alpha_k \in \phi$ constitute a basis of the GF(p)–vector space spanned by ϕ. Then $T_1 = \bigcap_{i=1}^{k} \ker(\alpha_i)$, which has codimension k in T:

$$(**) \qquad TR(L) = \dim T/C(L_p) \cap T = k + \dim T_1 - \dim C(L_p) \cap T .$$

$T_0 \cap T_1$ is contained in K_p and centralizes K. Then $T_0 \cap T_1 \subset C(K_p)$ and we obtain

$$(***) \qquad \dim C(K_p) \cap T_0 \ge \dim (T_0 \cap T_1) .$$

Combining (*), (**), (***) one gets

$$\begin{aligned} k + \dim T_1 - \dim C(L_p) \cap T = TR(L) \\ \ge \dim T_0 + \dim T_1 - \dim (T_0 \cap T_1) - \dim C(L_p) \cap T \end{aligned}$$

so that

$$k \ge \dim T_0 - \dim (T_0 \cap T_1) \ge \dim T_0/C(K_p) \cap T_0 .$$

This is the result. □

§3. Nilpotent subalgebras

In the classification theory [3] tori of maximal dimension play an important role. Let L be simple and restricted and suppose that T is a torus of maximal dimension. Then $C_L(T)$ is a CSA and $C_L(T)$ is triangulable [8,9] (see below for definition). Moreover, any torus of a k–section with respect to T has toral rank of most k.

Very surprisingly, things hardly change if we consider non-restricted Lie algebras using the concepts of "p-envelope" and "absolute toral rank". In the following let L denote a Lie algebra with finite dimensional p-envelope L_p. If K is a subalgebra of L, then K_p denotes the p-subalgebra of L_p generated by K.

Lemma 3.1: Let $T \subset T_0 \subset L_p$ be tori of L_p. The following are equivalent:

a) $TR(T_0,L_p) = TR(T,L_p)$

b) $T_0 + C(L_p) = T + C(L_p)$.

Proof: Since $T \subset T_0$ we have $\dim T_0 + C(L_p)/C(L_p) = \dim T + C(L_p)/C(L_p)$ if and only if $T_0 + C(L_p) = T + C(L_p)$. □

Proposition 3.2: Let K be a subalgebra of L and $T \subset K_p \subset L_p$ a torus which is either a maximal torus of K_p or of toral rank $TR(T,L_p) = TR(K,L)$. Then $C_K(T)$ is a nilpotent subalgebra.

Proof: If T is a maximal torus of K_p then it is well-known that $C_{K_p}(T)$ is a CSA of K_p. Therefore $C_K(T) \subset C_{K_p}(T)$ is nilpotent. Next assume that T is a torus of maximal toral rank and $T_0 \subset K_p$ a maximal torus of K_p containing T. The maximality of the toral rank implies that $TR(T_0,L_p) = TR(T,L_p)$. Applying (3.1) we obtain $T_0 + C(L_p) = T + C(L_p)$ and hence $C_K(T_0) = C_K(T)$. The first part of the proof shows that $C_K(T_0)$ is nilpotent. □

Note that in general $C_K(T)$ is not a CSA of K. The following theorem yields a characterization of CSAs.

Proposition 3.3: Let $T \subset L_p$ be a maximal torus or a torus of maximal toral rank. The following conditions are equivalent:

a) $TR(C_L(T),L) = TR(T,L_p)$.

b) $T \subset C_L(T)_p + C(L_p)$, where $C_L(T)_p$ denotes the p-algebra in L_p generated by $C_L(T)$.

c) $C_L(T)$ is a CSA.

Proof: $C_L(T)$ is a nilpotent algebra by Proposition 3.2. Then $C_L(T)_p$ is nilpotent, too, and has a unique maximal torus T_1. $T_1 + T$ is a torus. If T is a torus of maximal toral rank then $TR(T,L_p) = TR(T_1+T,L_p)$. Lemma 3.1 implies that $T + C(L_p) = T_1 + T + C(L_p)$ and hence $C_L(T) = C_L(T_1+T)$. Thus we may assume that $T_1 \subset T$ in either case.

a) ⇒ b): Since $TR(C_L(T),L) = TR(T_1,L_p)$, our present assumption a) and Lemma 3.1 show that $T_1 + C(L_p) = T + C(L_p)$. This yields b).

b) ⇒ c): We have to prove that

$$C_L(T) \supset \mathrm{Nor}_L(C_L(T)) := \{x \in L \mid [x,y] \in C_L(T)\ \forall y \in C_L(T)\}\ .$$

Take any $x \in \mathrm{Nor}_L(C_L(T))$. Then

$$[x,C_L(T)_p)] \subset \sum_{i\geq 0} (\mathrm{ad}\ C_L(T))^{p^i}(x) \subset C_L(T)\ .$$

Since we are assuming $T \subseteq C_L(T)_p + C(L_p)$, we have $[x,T] \subseteq C_L(T)$. Therefore

$$[T,x] = [T^{[p]},x] \subset (\mathrm{ad}\ T)^p(x) = 0$$

and $x \in C_L(T)$.

c) ⇒ a): Remember our assumption $T_1 \subset T$ where T_1 is the unique maximal torus of $C_L(T)_p$. We have to prove that $TR(T_1,L_p) = TR(T,L_p)$, i.e.

$$T_1 + C(L_p) = T + C(L_p)\ .$$

Assume on the contrary that $T_1 + C(L_p) \subsetneqq T + C(L_p)$. Decompose $L_p = \sum_{\alpha\in\Delta} L_{p,\alpha}(T)$ into eigenspaces with respect to T. The above assumption means that there is some $\beta \neq 0$ such that $\beta(T_1) = 0$, $\beta(T) \neq 0$. Therefore

$$0 \neq L_{p,\beta}(T) = [T,L_{p,\beta}] \subset L_p^{(1)} \subset L$$

and hence $L_{p,\beta}(T) \subset C_L(T_1)$. Thus we have

$$C_L(T_1) \supsetneqq C_L(T)\ .$$

Therefore $C_L(T_1)/C_L(T)$ is a nonzero $C_L(T)$–module, which is a restricted module on which T_1 acts nilpotently. According to an earlier remark in Section 1, $C_L(T)$ acts nilpotently on it. Engel's theorem now shows that there is $x \in C_L(T_1)\setminus C_L(T)$ with $[x,C_L(T)] \subset C_L(T)$. Then $C_L(T)$ is not a CSA. □

<u>Theorem 3.4</u>: Let $T \subset L_p$ be a torus of maximal toral rank. Suppose that K is a k–section with respect to T. If T_0 denotes a torus of K_p of toral rank k in K_p, then $C_K(T_0)$ is contained in a CSA of L_p.

<u>Proof</u>: Let R be the maximal torus of $C(L_p)$. Then $T + R$ is a torus with $TR(T+R,L_p) = TR(T,L_p)$ and K is a k–section with respect to $T + R$. Hence we may assume that $R \subset T$. K is a k–section and as such it can be written $K = \sum_{\alpha\in\phi} L_\alpha(T)$. Let $\alpha_1,...,\alpha_k \in \phi \subset T^*$ constitute a basis of the GF(p)–vector space spanned by ϕ. Put $T_1 := \cap_{\alpha\in\phi} \ker(\alpha)$ and $T_2 := T_0 + T_1$.

a) $\dim T_0 = k + \dim T_0 \cap C(K_p)$ is true according to the assumption on the toral rank of T_0.

b) $\dim T_1 = \dim T - k$, as $T_1 = \cap_{\alpha\in\phi} \ker(\alpha) = \cap_{i=1}^{k} \ker(\alpha_i)$.

c) $T_0 \cap T_1 \subset T_0 \cap C(K_p)$, since T_1 centralizes K_p and T_0 is contained in K_p.

Since $T_2 \cap C(L_p) \subset R \subset T \cap C(L_p)$, we conclude,

$$\begin{aligned} TR(T_2,L_p) &= \dim T_2/T_2 \cap C(L_p) = \dim T_0 + \dim T_1 - \dim T_0 \cap T_1 - \dim T_2 \cap C(L_p) \\ &\geq (k + \dim T_0 \cap C(K_p)) + (\dim T - k) - \dim T_0 \cap C(K_p) - \dim T \cap C(L_p) \\ &= \dim T - \dim T \cap C(L_p) = TR(T,L_p) . \end{aligned}$$

The maximality of $TR(T,L_p)$ implies that $TR(T_2,L_p) = TR(T,L_p)$. Take any maximal torus T_3 of L_p containing T_2. Then $TR(T_3,L_p) = TR(T,L_p) = TR(T_2,L_p)$. Lemma 3.1 proves that $T_2 + C(L_p) = T_3 + C(L_p)$, and $C_{L_p}(T_2) = C_{L_p}(T_3)$ is a CSA. On the other hand

$$C_{L_p}(T_2) \supset C_K(T_2) = C_K(T_0) .$$ □

A subalgebra K of L is said to be <u>triangulable</u> if $K^{(1)}$ acts nilpotently on L. We also say that <u>K acts triangulably on L</u> in this case. Then $ad_L K$ can be represented simultaneously by upper triangular matrices.

The following is an improvement of Wilson's results [8,9].

Theorem 3.5: Let L be a Lie algebra with p–envelope L_p and T any torus of L_p. Assume that the characteristic of F is bigger than 7. If hypothesis 1) or 2) below holds then $H = C_L(T)$ is triangulable.

1) a) H is a CSA of L.

b) L has toral rank 1 with respect to H.

c) $L^{(1)} = L$.

2) L is simple and H is nilpotent.

The proof is given by a careful application of Wilson's arguments.

i) In both cases H and H_p are nilpotent. Let T_0 denote the unique maximal torus of H_p. Then $T+T_0$ is a torus and $C_L(T+T_0) = H$. Thus we may assume $T_0 \subset T$. Let $L = \sum_{\alpha\in\Phi} L_\alpha$ be the root space decomposition with respect to T. We consider as usual $\alpha \in \Phi$ a mapping on $H_p + T$ defined by

$$\alpha(h)^{p^r} := \alpha(h^{[p]^r})$$

for suitably big r. Then $\alpha(h)$ is the unique eigenvalue of ad h on L_α.

ii) Let Q be any subalgebra of $H_p + T$ with $\alpha(Q^{(1)}) = 0$. Then Q has a common eigenvector in L_α and therefore α is linear on Q. Hence there exists a unique maximal ideal I of H_p such that $\alpha(I) = 0$ for all $\alpha \in \Phi$. I is a p–ideal in H_p.

iii) Since H_p is nilpotent, Lemma 3.1 of [8] remains true (substituting T by T_0).

iv) In order to prove [8,3.2] we take an element $b \in H_p$ as is described in [8,3.1] and we assume that $\alpha([b,H]) = 0 \ \forall \alpha \in \Phi$. Since $[b,H]$ is by construction an ideal of H_p, the definition of I shows $[b,H] \subset I$, contradicting [8,3.1].

v) Proof of Proposition 3.3 of [8]: In case 1) of our theorem we observe that

$$L_0 = H,\ \ \Phi = GF(p)\alpha,\ \ H = H^{(1)} + \Sigma[L_{i\alpha}, L_{-i\alpha}]\,.$$

Then $H = \Sigma[L_{i\alpha}, L_{-i\alpha}]$. This proves the assertion.

In case 2) of our theorem the proof in [8] works without any changes.

vi) The subsequent proof of the theorem in [8] for the first case does not require simplicity but just Proposition 3.3. We are done in that case.

Consider case 2). We have an element $b \in H_p$ as it is described in [8,3.1] and distinguished roots $\alpha,\beta \in \Phi$ with $\alpha([b,[L_\beta,L_{-\beta}]]) \neq 0$, $\beta([b,[L_\alpha,L_{-\alpha}]]) \neq 0$, $\alpha([b,[L_\alpha,L_{-\alpha}]]) = 0$, $\beta([b,[L_\beta,L_{-\beta}]]) = 0$. Then H is a CSA in

$$L(\alpha,\beta) := \sum_{i,j\in GF(p)} L_{i\alpha+j\beta}\,.$$

The arguments in [9] now yield the result.

vii) As a consequence, we have $H_p = T_0+I$. Then $H^{(1)} \subset I$. Since all roots vanish on I, all its elements act nilpotently on L. Then H is triangulable. □

Theorem 3.5 shows that in the situation of Theorem 3.4 $C_K(T_0)$ acts triangulably on K, whenever L is simple.

§4. Lie algebras of small absolute toral rank.

In this section we consider Lie algebras L with $TR(L) \leq 2$. For the sake of simplicity we assume throughout this section $char(F) > 7$ although sometimes a weaker hypothesis is sufficient.

Theorem 4.1: Let K be a subalgebra of L. Put $K^\infty := \bigcap_{n\geq 1} K^n$.

1) K acts nilpotently on L if and only if $TR(K,L) = 0$.

2) Suppose that $TR(K,L) = 1$. Then K is solvable if and only if K^∞ acts nilpotently on L.

3) Assume that $TR(K,L) = 1$ and that K is not solvable. Then

   a) $rad(K)$ and $rad(K^\infty)$ act nilpotently on L.

   b) $K^\infty/rad(K^\infty) \in \{s\ell(2), W(1;\underline{1}), H(2;\underline{1})^{(2)}\}$.

Proof: 1) K acts nilpotently on $L \Leftrightarrow K_p$ acts nilpotently on $L_p \Leftrightarrow$ every x in K_p acts nilpotently on L_p. Thus if K acts nilpotently on L every torus $T \subset K_p$ acts nilpotently on L_p. This, however, implies $T \subset C(L_p)$ and $TR(T,L_p) = 0$. On the other hand let x be an arbitrary element of K_p and $t = x^{[p]^r}$ its semisimple part. The assumption $TR(K,L) = 0$ means $TR(T,L_p) = 0$ for every torus $T \subset K_p$. Since t is a member of some torus T this implies that $t \in C(L_p)$ and $(ad\, x)^{p^r} = ad\, x^{[p]^r} = 0$.

2) If K^∞ acts nilpotently on L, then in particular K^∞ is nilpotent. As K/K^∞ is nilpotent, K is

solvable. Assume that K^∞ does not act nilpotently on L. Then there exists $h \in K^\infty$ which does not act nilpotently on L. Decompose $K = \sum_{\alpha\in\Delta} K_\alpha(\operatorname{ad} h)$ with respect to $\operatorname{ad} h$ and put $H := K_0(\operatorname{ad} h)$. Let $t := h^{[p]^r}$ be the semisimple part of h in K_p, T_0 the torus generated by t and T a maximal torus of K_p containing T_0. Our present assumption $TR(K,L) = 1$ implies

$$1 \geq TR(T,L_p) \geq TR(T_0,L_p) \geq 1 ,$$

hence $TR(T_0,L_p) = TR(T,L_p) = 1$. According to Proposition 3.2 we have that $H = K_0(\operatorname{ad} h) = C_K(T_0)$ is nilpotent.

$$I := \sum_{\alpha\neq 0} K_\alpha + \sum_{\alpha\neq 0} [K_\alpha,K_{-\alpha}] \text{ is an ideal of } K.$$

Remember that $h \in K^\infty$. Therefore $K_\alpha = [K_\alpha,h] \subset K^\infty$ is true for $\alpha \neq 0$ and I has the properties

$$K = H + I,\ I \subset K^\infty .$$

The nilpotency of H implies that $K^\infty \subset I$, proving $K^\infty = I$ and in particular

$$h \in K^\infty \cap H = \sum_{\alpha\neq 0} [K_\alpha,K_{-\alpha}] .$$

Then K^∞ and K are not solvable.

3) a) K is not solvable and therefore $K^\infty/\operatorname{rad}(K^\infty)$ is not nilpotent. From the inequalities of Section 2 we have

$$1 = TR(K,L) \geq TR(K^\infty,L) \geq TR(K^\infty/\operatorname{rad}(K^\infty)) + TR(\operatorname{rad}(K^\infty),L) .$$

This in combination with $TR(K^\infty/\operatorname{rad}(K^\infty)) \geq 1$, (which follows from the first part of this theorem since $K^\infty/\operatorname{rad}(K^\infty)$ is not nilpotent), prove that $TR(\operatorname{rad}(K^\infty),L) = 0$. Then, $\operatorname{rad}(K^\infty)$ acts nilpotently on L. A similar reasoning yields that $\operatorname{rad}(K)$ acts nilpotently on L.

b) The assumption $TR(K,L) = 1$ and the nonnilpotency of K implies $TR(K) = 1$. Then K has toral rank 1 with respect to any CSA. From [10,6] it follows that $K^\infty/\operatorname{rad}(K^\infty)$ is simple and one of $s\ell(2)$, $W(1;\underline{n})$, $H(2;\underline{n};\phi)^{(2)}$. Moreover, the assumption $TR(K^\infty/\operatorname{rad}(K^\infty)) = 1$ now implies that it is one of $s\ell(2)$, $W(1;\underline{1})$, $H(2;\underline{1})^{(2)}$ [3,(2.2.3)]. □

In the context of classification it is natural to assume that CSAs act triangulably.

Corollary 4.2. Let L be a Lie algebra with $TR(L) = 1$. Assume that L has a CSA, which is triangulable. Then

1) L is solvable if and only if $L^{(1)}$ is nilpotent.

2) If L is nonsolvable, one of the following cases occurs

   a) $L/rad(L) \cong s\ell(2)$

   b) $L/rad(L) \cong W(1;\underline{1})$

   c) $H(2;\underline{1})^{(2)} \subset L/rad(L) \subset H(2;\underline{1})$.

$L/rad(L)$ is restrictable.

Proof: Let H be a triangulable CSA of L and T the unique maximal torus of H_p in L_p. Decompose $L = \sum_{i \in GF(p)} L_{i\alpha}$ with respect to H. (Note that this means a decomposition with respect to T; now observe that $\dim T/T \cap C(L_p) = 1$.) Let S be any ideal of L with decomposition $S = \sum_{i \neq 0} S \cap L_{i\alpha} + S \cap H$. Take any $x \in \bigcup_{i \neq 0} S \cap L_{i\alpha}$ with semisimple part $t := x^{[p]^r}$. Then $[t,T] = 0$ and $T + Ft$ is contained in a torus T_1. Our assumption on the toral rank implies that

$$t \in T_1 \subset T + C(L_p).$$

On the other hand,

$$\alpha(t)x = [t,x] = (\mathrm{ad}\, x)^{p^r}(x) = 0 .$$

Hence $\alpha(t) = 0$ and $t \in C(L_p)$. This means that every element of $\bigcup_{i \neq 0} S \cap L_{i\alpha}$ acts nilpotently on L. As a result we obtain that S acts nilpotently on L if and only if $S \cap H$ does. Moreover, if $S \cap H$ does not act nilpotently, then $S+H = L$.

Apply this to $S_1 := \sum_{i \neq 0} L_{i\alpha} + \sum_{i \neq 0} [L_{i\alpha}, L_{-i\alpha}]$ and $S_1 + H^{(1)} = L^{(1)}$. If $L^{(1)}$ is not nilpotent, then $L^{(1)} \cap H = H^{(1)} + S_1 \cap H$ acts nonnilpotently. Since H acts triangulably, $S_1 \cap H$ acts nonnilpotently. Then $S_1^{(1)} = S_1 \neq 0$ and L is not solvable. This proves 1). In order to prove 2) we assume that L is nonsolvable and hence that $S_1 \cap H$ acts nonnilpotently.

Let S' be an ideal of S_1 with $S_1 \cap rad(L) \subset S'$. It is easy to see that

$$S := \sum_{j \geq 0} (\mathrm{ad}\, H)^j(S')$$

is an ideal of L (observe $S_1+H = L$). Note that $S\cap H \subset S'\cap H+H^{(1)}$. Thus if $S'\cap H$ acts nilpotently on L, then so does $S\cap H$ and S is nilpotent. Therefore $S' \subset S \subset rad(L)$ in this case. If $S'\cap H$ does not act nilpotently, then $S_1 \subset S' + H$ and therefore $S_1 = S'$. As a result, $G := S_1/S_1 \cap rad\, L$ is simple and $L/rad\,(L)$ acts on it faithfully by derivations. As in Theorem 4.1 either $G = s\ell(2)$ or $G = W(1;\underline{1})$ (which implies that $L/rad\,(L) = G$) or $H(2;\underline{1})^{(2)} = G$. Then
$L/rad\,(L) \to Der(H(2;\underline{1})^{(2)})$ and, in addition, $L/rad(L)$ has absolute toral rank 1. The structure of $Der(H(2;\underline{1})^{(2)})$ yields that $L/rad(L) \subset H(2;\underline{1})$. $L/rad(L)$ carries a p-structure in any case. □

We now consider Lie algebras with absolute toral rank 2. From the viewpoint of classification theory Lie algebras L having a CSA of toral rank 1 in L may be considered as well known. One of the major problems which arises by considering tori in some p-envelope L_p of L (rather than in a restricted Lie algebra L) is, whether the centralizer in L of a maximal torus is a CSA of L or not. The centralizer in L_p is of course a CSA of L_p, but this information is not sufficient.

It is also false in general that every CSA of L is the centralizer of a maximal torus of L_p.

<u>Theorem 4.3</u>. Let L be a Lie algebra with p-envelope L_p and T a maximal torus of L_p or a torus of maximal toral rank. Assume that $TR(L) = 2$. Then either L has toral rank 1 with respect to some CSA, which is a section with respect to T, or $C_L(T)$ is a CSA of L and $TR(T,L_p) = 2$.

<u>Proof</u>: Put $H := C_L(T)$. H and H_p are nilpotent by Proposition 3.2 and therefore H_p has a unique maximal torus T_0.

If $T \subset T_0 + C(L_p)$, then H is a CSA and $TR(T_0,L_p) = TR(T,L_p)$ by Proposition 3.3. We are done if $TR(T,L_p) = 2$. If $TR(T,L_p) = 1$ then $TR(H,L) = 1$ and H is a 0-section with respect to T.

Next consider the case that $T \not\subset T_0 + C(L_p)$. Note that $[T,T_0] = 0$. Therefore $T + T_0$ is a torus. If T is maximal, then $T_0 \subset T$. If T has maximal toral rank, then let T' be a maximal torus of L_p containing $T + T_0$. We have by the maximality condition

$$TR(T',L_p) = TR(T,L_p)$$

and hence

$$T' + C(L_p) = T + C(L_p) .$$

Substituting T by T' we may assume that in either case T is a maximal torus containing T_0.

a) Our present assumption implies that $T_0 \not\equiv T \bmod (C(L_p))$. Decompose $L = \sum_{\alpha\in\Delta} L_\alpha$ with respect to T. There exists $\beta \in \Delta$ such that

$$\beta(T_0) = 0, \ \beta(T) \neq 0 .$$

For every $x \in \bigcup_{i\in GF(p)\setminus 0} L_{i\beta}$ let $t(x) := x^{[p]^r}$ be the semisimple part of x. Note that $[t(x),T] = 0$. The maximality of T implies (as $T + Ft(x)$ is contained in some torus) $t(x) \in T$.

Because of

$$\beta(t(x))x = [t(x),x] = (\operatorname{ad} x)^{p^r}(x) = 0$$

we obtain

$$t(x) \in \ker(\beta) \quad \forall\, x \in \bigcup_{i\in GF(p)\setminus 0} L_{i\beta} .$$

Put $H_1^\beta := \sum_{i\in GF(p)} L_{i\beta}$. Since $\beta(T_0) = 0$ the above result shows that $\bigcup_{i\in GF(p)} L_{i\beta}$ is a Lie set acting nilpotently on H_1^β. Then H_1^β is a nilpotent subalgebra. Let $T_1 = T_1^\beta$ be the unique maximal torus of $(H_1^\beta)_p$

$$[T_1,T] = [T_1^{[p]},T] \subset (\operatorname{ad} T_1)^p(T) \subset (\operatorname{ad} T_1)^{p-1}([(H_1)_p,T]) \subset (\operatorname{ad} T_1)^{p-1}((H_1)_p) = 0 .$$

Therefore $T_1 + T$ is a torus and the maximality of T implies

$$T \supset T_1^\beta \supset T_0 .$$

b) If $T_1^\beta \subset C(L_p)$ then $T_0 \subset C(L_p)$ and $C_L(T)$ acts nilpotently on L. In this case a) applies for all roots $\beta \neq 0$.

If $\bigcup_{\beta\in\Delta} T_1^\beta \subset C(L_p)$ then $\bigcup_{\beta\in\Delta} H_1^\beta$ acts nilpotently on L. The Engel–Jacobson theorem proves that L is nilpotent, contradicting the assumption $TR(L) \neq 0$. Thus there is some γ with $\gamma(T_0) = 0$, $\gamma(T) \neq 0$, $T_1^\gamma \not\subset C(L_p)$.

c) According to b) we may take $\beta \in \Delta$ such that $T_1^\beta =: T_1 \not\subset C(L_p)$. Then

$$0 \neq \dim T_1/T_1\cap C(L_p) \leq \dim T\cap\ker(\beta)/T\cap\ker(\beta)\cap C(L_p) \leq 1 .$$

Therefore

$$\mathrm{TR}(H_1^\beta,L) = \dim T_1/T_1 \cap C(L_p) = 1$$

$$\dim T/T\cap C(L_p) = 2$$

$$T_1 \equiv T \cap \ker(\beta) \ \ \mathrm{mod}(C(L_p)) .$$

The last congruence implies

$$C_L(T_1) = \sum_{i\in GF(p)} L_{i\beta} = H_1^\beta .$$

Since $\mathrm{TR}(H_1^\beta,L) = \mathrm{TR}(T_1,L_p) = 1$, H_1^β is a CSA of L by Proposition 3.3; it is a 1–section with respect to T. This proves the theorem. □

Theorem 4.4. Assume that $\mathrm{TR}(L) = 2$. Suppose that H is a CSA of L. Then either L has toral rank 1 with respect to H or there is a maximal torus T of L_p of toral rank 2 such that $C_L(T) = H$.

Proof: Let T_0 denote the unique maximal torus of H_p and T a maximal torus of L_p containing T_0. If the toral rank of T_0 in L_p is 0, then H acts nilpotently on L. This means, as H is a CSA, that $L = H$ is nilpotent. Then, however, $\mathrm{TR}(L) = 0$, a contradiction.

If the toral rank of T_0 is one, then L has toral rank 1 with respect to H.

If the toral rank of T_0 is 2 (note that it cannot be bigger since $\mathrm{TR}(L) = 2$) then $T \equiv T_0 \ \mathrm{mod}(C(L_p))$. Therefore $C_L(T) = C_L(T_0) = H$. □

Corollary 4.5. Suppose that $\mathrm{TR}(L) = 2$ and $\mathrm{TR}(H,L) = 2$ for every CSA of L. Then there is a one–to–one correspondence between maximal tori of L_p and CSAs of L, by which every maximal torus T is mapped to $C_L(T)$, and to any CSA H is associated the unique maximal torus of $H_p+C(L_p)$.

Proof: Let T be a maximal torus of L_p and $H := C_L(T)$. Theorem 4.3 proves that H is a CSA. Let T_0 denote the unique maximal torus of H_p. Our present assumption yields $\mathrm{TR}(T_0,L_p) = \mathrm{TR}(T,L_p) = 2$ and hence $T \equiv T_0 \ \mathrm{mod}(C(L_p))$ by Lemma 3.1. If T_1 denotes the maximal torus of $C(L_p)$, the maximality of T implies $T = T_0+T_1$.

If, on the other hand, H is any CSA of L, Theorem 4.4 yields the existence of a maximal torus T of L_p of toral rank 2 with $H = C_L(T)$. The above considerations show that T is uniquely determined by H. □

As a result of these theorems we obtain a far-reaching reduction: Every algebra L with $TR(L) \leq 2$ is either nilpotent (which is a very inconvenient case and has to be ruled out separately) or of toral rank 1 with respect to some CSA (L then is considered as well known) or there is a one-to-one correspondence between maximal tori of L_p and CSAs of L. This result enables us to apply some methods which have been developed by Block–Wilson [3] and Benkart–Osborn [1] in order to get some classification results.

We restate all this and some consequences.

Corollary 4.6. Assume that $TR(L) \leq 2$. Then one of the following cases occur:

(1) $TR(L) = 0$, L is nilpotent.

(2) L has a CSA H such that $TR(H,L) = 1$.

(3) a) Every maximal torus of L_p has toral rank 2.

b) There is a one-to-one correspondence between maximal tori of L_p and CSAs of L.

c) Suppose that T is a maximal torus of L_p and $H := C_L(T)$ the corresponding CSA of L. Then every H-invariant subspace of L is T-invariant. In particular, every subalgebra of L containing H and any ideal of L decomposes into root spaces with respect to T.

Proof: Let H be a CSA of L. If $TR(H,L) = 0$ then H acts nilpotently on L by Theorem 4.1, proving that $H = L$ is nilpotent. If $TR(H,L) = 1$ we are in case (2). Hence if (1) and (2) are not true then (3a) happens to hold. Corollary 4.5 implies (3b). Every H-invariant subspace is invariant under $H_p+C(L_p)$ and hence under $T \subset H_p+C(L_p)$. □

Since we want to describe the structure of L we are now mainly interested in the third case of Corollary 4.6 and state this as an assumption:

$$\text{(A)}\quad \begin{array}{l} TR(L) = 2 \\ TR(H,L) = 2 \text{ for every CSA } H \text{ of } L. \end{array}$$

Corollary 4.6 shows that under the assumption $TR(L) \leq 2$ (A) is equivalent to (3) of that corollary.

For any root space decomposition $L = \sum_{\alpha \in \Delta} L_\alpha$ with respect to some CSA or torus we denote by $L^{(\beta)} := \sum_{i \in GF(p)} L_{i\beta}$ the 1–section determined by a root β.

Let G be a subalgebra of L. By $R(G)$ (or more explicitly $R(G,L)$) we denote the maximal ideal of G which acts nilpotently on L.

Note that $R(G) \subset rad(G)$.

Theorem 4.7. Assume that $TR(L) \leq 2$. Suppose that G is a subalgebra of L. Then either

(1) There is some CSA H of G such that

$$G = G^{(\beta)} + R(G)$$

for some root β with respect to H ($\beta = 0$ is allowed) or

(2) $R(G) = rad(G)$ and G as well as $G/rad(G)$ satisfy (A).

Proof: We assume that (1) is not true and prove that in this case (2) holds. Corollary 4.6 applied to G shows that G satisfies (A).

Consider the canonical homomorphism

$$\pi\colon G \to G/R(G) .$$

Let H' be a CSA of $G/R(G)$ and H a CSA of $\pi^{-1}(H')$. Then H is a CSA of G. If $TR(H',G/R(G)) = 1$, then G decomposes with respect to H as $G = \sum_{i \in GF(p)} G_{i\beta}(H)+R(G)$. Since this contradicts our assumption, $G/R(G)$ satisfies (A). Next assume, that J' is a nonzero abelian ideal of $G/R(G)$. Put $J := \pi^{-1}(J')$. Since $J \not\subset R(G)$ we have $TR(J,L) \neq 0$ by Theorem 4.1. Hence Proposition 2.3 yields the contradiction

$$2 \geq TR(L) \geq TR(G,L) \geq TR(G/J) + TR(J,L) \geq 2+1 .$$

Therefore $G/R(G)$ is semisimple and $R(G) = rad(G)$. □

Applying this theorem to $G = L$ either we obtain some structural insight into L (case (1)) or we are led to the consideration of $L/rad(L)$.

Let $A(n;\underline{1})$ denote the truncated polynomial ring $F[x_1,\dots,x_n]/(x_1^p,\dots,x_n^p)$ and $W(n;\underline{1}) :=$

Der(A(n;$\underline{1}$)) the n–th Jacobson–Witt algebra.

Theorem 4.8. Assume that

a) L is semisimple

b) L satisfies (A).

c) $C_L(T)$ is triangulable for every torus $T \subset L_p$ with maximal toral rank in L_p.

Then one of the following cases occur:

(1) There exist ideals S_1, S_2 of L, which are simple as algebras and satisfy $TR(S_i) = 1$ $(i = 1,2)$, such that

$$S_1 \oplus S_2 \subset L \subset L_p \subset Der(S_1) \oplus Der(S_2) .$$

(2) There exists a subalgebra $S \subset L$, which is simple as an algebra and satisfies $TR(S) = 1$, such that

$$S \otimes A(n;\underline{1}) \subset L \subset L_p \subset Der(S) \otimes A(n;\underline{1}) + F \otimes W(n;\underline{1})$$

for some $n \geq 0$.

(3) There exists a subalgebra $S \subset L$, which is simple as an algebra and satisfies $TR(S) = 2$, and $n \geq 0$ such that

$$S \otimes A(n;\underline{1}) \subset L \subset S \otimes A(n;\underline{1}) + H$$

for every CSA H of L.

If $n \neq 0$ then $TR(H_0,L) = 1$ for every CSA H_0 of S.

Proof: Let $G := S_1 \oplus \cdots \oplus S_r$ be the socle of the semisimple algebra L, i.e. the sum of all minimal ideals of L. Since none of them is nilpotent, we obtain

$$2 = TR(L) \geq TR(G) = \sum_{i=1}^{r} TR(S_i) \geq r .$$

Hence either $r = 2$, $G = S_1 \oplus S_2$, $TR(S_i) = 1$ $(i = 1,2)$ or $r = 1$, $G = S_1$, $TR(G) \leq 2$ is true.

We make some preliminary observations.

(a) Assume first, that G has a CSA H such that $TR(H,G) = 2$. Let T_0 be the maximal torus of H_p. As $TR(T_0,G_p) = 2 = TR(T_0,L_p)$, $H_1 := C_L(T_0)$ acts triangulably on L. Since $T_0 \subset G_p$, we have

$$L = [T_0,L] + C_L(T_0) \subset G + H_1 .$$

Let $J \neq 0$ be an ideal of G. It is easily seen, that

$$J' := \sum_{j\geq 0} (\mathrm{ad}\, H_1)^i(J) \subset \sum_{\alpha\neq 0} G_\alpha(T_0) + C_J(T_0) + H_1^{(1)}$$

is an ideal of L which is contained in G.

(a1) If G is a minimal ideal of L (i.e. $r = 1$) then $J' = G$, hence

$$H \subset C_G(T_0) \subset C_J(T_0) + H_1^{(1)} .$$

$H_1^{(1)}$ is an ideal in $C_J(T_0) + H_1^{(1)}$. Therefore $(C_J(T_0))_p + (H_1^{(1)})_p$ is the p–envelope of $C_J(T_0) + H_1^{(1)}$ in L_p. It contains $T_0 \subset H_p$. We have for $\alpha \neq 0$

$$G_\alpha(T_0) = [T_0,G_\alpha] \subset [J_p,G_\alpha] + [(H_1^{(1)})_p,G_\alpha] \subset J + [(H_1^{(1)})_p,G_\alpha] .$$

Since $H_1^{(1)}$ acts nilpotently this yields

$$\sum_{\alpha\neq 0} G_\alpha \subset J ,$$

while the minimality of G implies

$$G = \sum_{\alpha\neq 0} G_\alpha + \sum_{\alpha\neq 0} [G_\alpha,G_{-\alpha}] \subset J .$$

Therefore G is simple.

(a2) If $r = 2$ and $G = S_1 \oplus S_2$ we assume that $J \subset S_1$. Since $TR(S_1) = 1$ we have from Proposition 2.3 that either $TR(S_1/J) = 0$ or $TR(J,S_1) = 0$. In the first case S_1/J is nilpotent, and as $S_1^{(1)} = S_1$ this means $S_1 = J$. In the latter case J acts nilpotently on S_1 by Theorem 4.1. Then J' is an ideal of L being contained in S_1 (i.e. $J' = S_1$) and

$$C_{J'}(T_0) \subset C_J(T_0) + H_1^{(1)} ,$$

which acts nilpotently on S_1. The first remark yields

$$C_{S_1}(T_0) = C_{J'}(T_0) ,$$

and since $H = C_G(T_0) = C_{S_1}(T_0) \oplus C_{S_2}(T_0)$ is a CSA of G, the second part gives $S_1 \subset H$. Then $S_1 \subset \mathrm{rad}(L) = 0$, a contradiction. Hence S_1,S_2 are simple.

b) Assume that $TR(H,G) = 1$ for every CSA of G. Note that $r = 1$ in this case. Then G

is a minimal ideal of L, i.e. L–irreducible. From [2] we conclude that for some simple subalgebra $S \subset G$ and some $n \geq 0$

$$G \cong S \otimes A(n;\underline{1}) .$$

c) L acts on G via "ad" as derivations

$$\varphi: L \longrightarrow \mathrm{Der}(G), \quad \varphi(x) := \mathrm{ad}_G x .$$

Since G is the socle of L and L is semisimple, the assumption $\ker(\varphi) \neq 0$ yields the contradiction

$$0 \neq \ker(\varphi) \cap G \subset \mathrm{rad}(L) = 0 .$$

Thus φ is injective and therefore Der(G) contains a p–envelope L_p of L

$$G \subset L \subset L_p \subset \mathrm{Der}(G) .$$

We now put pieces together. If $r = 2$ then $\mathrm{Der}(G) = \mathrm{Der}(S_1) \oplus \mathrm{Der}(S_2)$ and the above yields case (1). Suppose that $r = 1$. Then $G \cong S \otimes A(n;\underline{1})$ for some simple S and $n \geq 0$ (cf. (b)). Note that $\mathrm{Der}(S \otimes A(n;\underline{1})) = \mathrm{Der}(S) \otimes A(n;\underline{1}) + F \otimes W(n;\underline{1})$. Hence if $TR(S) = 1$ we are in case (2). Suppose that $TR(S) = 2$.

Let H be the CSA of L and T the maximal torus of H_p. Consider the canonical restricted homomorphism $\pi: L_p \longrightarrow L_p/G_p$. $\pi(T)$ is a torus of L_p/G_p. On the other hand, since L_p is centerless, we have the following inequalities

$$\begin{aligned} 2 = TR(L_p) = MT(L_p) &\geq MT(L_p/G_p) + MT(G_p) \\ &\geq MT(L_p/G_p) + TR(S) = MT(L_p/G_p) + 2 \end{aligned}$$

which prove that $MT(L_p/G_p) = 0$. Therefore $\pi(T) = 0$ and hence $T \subset G_p$. This implies

$$L = [T,L] + C_L(T) \subset [G,L] + C_L(T) \subset G + C_L(T) .$$

Condition (A) means that Corollary 4.6(3) applies to L. Therefore $C_L(T) = H$ and $L \subset G + H$.

Let H_0 be a CSA of S and T_0 the maximal torus of $(H_0)_p$. If $TR(T_0,L_p) = 2$ then $H_1 := C_G(T_0)$ is nilpotent by Proposition 3.2. Moreover, $H_0 \subset H_1$ and therefore $\mathrm{Nor}_G(H_1) \subset H_1$. Thus H_1 is a CSA of G with $TR(H_1,G) = 2$. (a1) of this proof shows that G is simple. As a result, $n \neq 0$ implies that $TR(H_0,L) = 1$ for every CSA of S. □

These results will be used for an attack on the general classification problem. We start with a simple Lie algebra L, take a torus T of maximal toral rank of some semisimple p–envelope L_p and consider 2–sections K with respect to T. Theorem 2.6 implies that $TR(K) \leq 2$. If $TR(K) \leq 1$ we can apply Theorem 4.1. Thus we may assume $TR(K) = 2$ and take any torus $T_0 \subset K_p$ of toral rank 2. Theorems 3.4 and 3.5 show that $C_K(T_0)$ acts triangulably on K. $R(K)$ is by definition the nilradical of K. Either $K/R(K)$ is of toral rank 1 with respect to some CSA, or $K/R(K)$ is semisimple and K as well as $K/R(K)$ satisfy (A) by Theorem 4.7. Consider the case that $\bar{K} := K/R(K)$ is semisimple and $K, \bar{K}$ satisfy (A). In order to prove that the assumptions of Theorem 4.8 are true for $\bar{K}$, we observe that due to (A) for K and $\bar{K}$, CSAs are in one–to–one correspondence with tori of maximal toral rank. Hence any CSA of K acts triangulably on K by Theorems 3.4 and 3.5. Since any CSA of $\bar{K}$ is the image of a CSA of K, the former acts triangulably on $\bar{K}$. Therefore Theorem 4.8 is applicable to $\bar{K}$. Again, either one can describe $\bar{K}$ in terms of subalgebras of toral rank 1 (with respect to some CSA) or (a subcase of Theorem 4.8(3)) there is a simple algebra S satisfying (A) such that $\bar{K} = S+H$ for every CSA H of $\bar{K}$.

Theorem 4.8 in combination with Corollary 4.2 is the analogue of [3, Theorem 4.1.1].

References

[1] G. Benkart, J. M. Osborn, Simple Lie algebras of characteristic p with dependent roots, to appear Trans. Amer. Math. Soc.

[2] R. E. Block, Differentiably simple algebras, Bull. Amer. Math. Soc. 74 (1968), 1086–1090.

[3] R. E. Block, R. L. Wilson, Classification of the restricted simple Lie algebras, J. Algebra 114 (1988) 115–259.

[4] R. Farnsteiner, H. Strade, On the derived length of solvable Lie algebras, to appear in Math. Ann.

[5] H. Strade, Einige Vereinfachungen in der Theorie der modularen Lie–Algebren, Abh. Math. Sem. Univ. Hamburg 54 (1984), 257–265.

[6] H. Strade, Cyclically graded algebras, in preparation.

[7] H. Strade, R. Farnsteiner, Modular Lie algebras and their representations, Marcel Dekker Textbooks and Monographs, Vol. 116 (1988).

[8] R. L. Wilson, Cartan subalgebras of simple Lie algebras, Trans. Amer. Math. Soc. 234 (1977), 435–446.

[9] R. L. Wilson, Correction to "Cartan subalgebras of simple Lie algebras", preprint.

[10] D. J. Winter, The structure of cyclic Lie algebras, Proc. Amer. Math. Soc. 100 (1987), 213–219.

1980 AMS subject classification (1985 revision) 17B50, 17B20.

Department of Mathematics
University of Hamburg
Bundesstrasse 55
2 Hamburg 13, FR Germany

and

Department of Mathematics
University of Wisconsin
Madison, WI 53706

This paper is in final form and no version of it will appear elsewhere.

DIFFERENTIAL FORMS AND THE ALGEBRA W(m:n)

Robert Lee Wilson*

Introduction.

The finite–dimensional simple Lie algebras of Cartan type ([5,6,15] cf. [3]) are defined by use of two types of subalgebras of the infinite–dimensional Lie algebra W(m): the infinite–dimensional algebras S(m), H(m), K(m) and the finite–dimensional algebras W(m:n). The usual definitions of these subalgebras are quite different. The algebras S(m), H(m), K(m) are defined in terms of appropriate differential forms while W(m:n) is defined as the stabilizer in W(m) of a finite–dimensional subalgebra $\mathfrak{A}(m:n) \subseteq \mathfrak{A}(m)$. For some purposes the definition of S(m), H(m), K(m) in terms of differential forms is much easier to work with than the definition of W(m:n).

In this paper we will show that the algebras W(m:n) may also be characterized in terms of differential forms and we will use this characterization to give a new proof of an important embedding theorem.

Our starting point is the fact that W(m:n) is isomorphic to a certain subalgebra of the Jacobson–Witt algebra W_n $(\mathbf{n} = (n_1,\dots,n_m),\ n = n_1 + \cdots + n_m)$ studied by Ree [9]. We will denote this subalgebra by $\mathcal{R}(m:\mathbf{n})$. Furthermore, there is a subalgebra $\mathcal{R}_*(m:\mathbf{n}) \supseteq \mathcal{R}(m:\mathbf{n})$ of W_n such that Der $W(m:\mathbf{n}) \cong \mathcal{R}_*(m:\mathbf{n})$ and $\mathcal{R}_*(m:\mathbf{n})$ may be characterized as the annihilator in W_n of a certain collection of differential forms (Proposition 2.3).

One of the essential steps in the proof of the Recognition Theorem for simple Lie algebras of classical and Cartan types ([5,6,15], cf. [3, Theorem 1.2.2]) is to show that if $L \subseteq W(m)$ and $X(m:\mathbf{n})^{(2)} \subseteq \mathrm{gr}\, L \subseteq X(m:\mathbf{n})$ (where X = W,S,H or K and the filtration of W(m) is given by $\deg x_i = 1 + \delta_{XK}\delta_{im}$), then there is some automorphism Φ of W(m) such that $\Phi L \subseteq W(m:\mathbf{n})$. Each of the existing treatments of this part of the Recognition Theorem contains a result designed for this purpose ([6, Theorem 0], [15, Proposition 5.1]). (Related theorems appear in [8] and [11, §3.5].) However the proof of [15, Proposition 5.1] is very involved and the proof of [6, Theorem 0] does not seem to be correct.

(*) The author gratefully acknowledges partial support from The Institute for Advanced Study and from National Science Foundation grant number DMS–8603151.

We prove here, using the characterization of W(m:n) by differential forms, a result (Theorem 5.3) which is adequate for this step in the proof of the Recognition Theorem.

We believe that our characterization of W(m:n) may also be useful in the determination of forms of W(m:n) over nonalgebraically closed fields, perhaps allowing a treatment parallel to that of the restricted simple Lie algebras of Cartan type in [10].

This paper is organized as follows. In Section 1 we define the subalgebras $\mathcal{R}$(m:n) and $\mathcal{R}_*$(m:n) of W_n and prove they are isomorphic to W(m:n) and Der W(m:n) respectively. In Section 2 we show that $\mathcal{R}_*$(m:n) is the annihilator of a certain collection of differential forms. Using this we prove (Sections 3,4) conjugacy theorems for certain subalgebras of W_n which lead (in Section 5) to the embedding theorem (Theorem 5.3) needed for the Recognition Theorem.

Throughout the paper **Z** will denote the ring of integers and F will denote an algebraically closed field of characteristic $p > 0$.

The author is indebted to S. Serconek for helpful discussions during the preparation of this paper.

§1. The algebras $\mathcal{R}$(m:n) and $\mathcal{R}_*$(m:n).

Let $\mathbf{n} = (n_1,\dots,n_m)$ be an m–tuple of positive integers and $n = n_1 + \cdots + n_m$. As usual B_n will denote the truncated polynomial algebra on n indeterminates and W_n will denote Der B_n. In this section we will define subalgebras $\mathcal{R}(m:\mathbf{n}) \subsetneq \mathcal{R}_*(m:\mathbf{n})$ of W_n such that $\mathcal{R}(m:\mathbf{n}) \cong W(m:\mathbf{n})$ and $\mathcal{R}_*(m:\mathbf{n}) \cong \mathrm{Der}\, W(m:\mathbf{n})$.

We first recall the definitions of W(m) and of W(m:n) ([7,12], cf. [11,§3.5]). We let A(m) denote the monoid (under addition) of all m–tuples of nonnegative integers. For $\alpha,\beta \in A(m)$ we define $\begin{bmatrix}\alpha\\ \beta\end{bmatrix} = \prod_i \begin{bmatrix}\alpha(i)\\ \beta(i)\end{bmatrix}$. Define $\varepsilon_i \in A(m)$ by $\varepsilon_i(j) = \delta_{ij}$. Let $\mathfrak{A}(m)$ denote the (commutative, associative) algebra consisting of all formal sums of the elements $\{x^\alpha \mid \alpha \in A(m)\}$ with multiplication given by

$$x^\alpha x^\beta = \begin{bmatrix}\alpha+\beta\\ \alpha\end{bmatrix} x^{\alpha+\beta} .$$

Let D_i denote the derivation of $\mathfrak{A}(m)$ defined by $D_i x^\alpha = x^{\alpha-\varepsilon_i}$ for all $\alpha \in A(m)$ (where $x^\beta = 0$ if $\beta \notin A(m)$). Then $W(m) = \Sigma_i \mathfrak{A}(m) D_i$ is a subalgebra of Der $\mathfrak{A}(m)$. For any m–tuple $\mathbf{r} = (r_1,\dots,r_m)$ of positive integers we may give $\mathfrak{A}(m)$ and W(m) the structures of topologically graded and filtered algebras by setting

$$\deg x_i = -\deg D_i = r_i , \quad 1 \le i \le m .$$

We write $\mathfrak{A}(m)_i$ and $W(m)_i$ for the i–th terms in the filtrations and $\mathfrak{A}(m)_{[i]}$ and $W(m)_{[i]}$ for the degree i components in the gradations.

Let $A(m:n) = \{\alpha \in A(m) \mid 0 \le \alpha(i) < p^{n_i}, 1 \le i \le m\}$, $\mathfrak{A}(m:n) = \operatorname{span}\{x^\alpha \mid \alpha \in A(m:n)\}$ and $W(m:n) = \Sigma_i \mathfrak{A}(m:n)D_i$. Then $\mathfrak{A}(m:n)$ (respectively, $W(m:n)$) is a graded and filtered subalgebra of $\mathfrak{A}(m)$ (respectively, $W(m)$).

Now let $I = \{(i,j) \mid 1 \le i \le m, 1 \le j \le n_i\}$ and $J = \{(i,j) \in I \mid j < n_i\}$. Thus $|I| = n$ and so

$$B_n \cong F[x_{ij} \mid (i,j) \in I]/(x_{ij}^p \mid (i,j) \in I) .$$

Write D_{ij} for the derivation of B_n induced by $\partial/\partial x_{ij}$. Set

$$E_i = \sum_{j=1}^{n_i} (-1)^{j+1} x_{i1}^{p-1} \cdots x_{i,j-1}^{p-1} D_{ij} .$$

It is known ([13, Lemma 4]) that $\operatorname{Der} W(m:n) = \operatorname{ad} W(m:n) + \operatorname{span}\{(\operatorname{ad} D_i)^{p^j} \mid (i,j) \in J\}$. Since ad is injective we may (and do) identify $\operatorname{Der} W(m:n)$ with $W(m:n) + \operatorname{span}\{D_i^{p^j} \mid (i,j) \in J\}$ $\subseteq \operatorname{Der} \mathfrak{A}(m:n)$.

<u>Definition 1.1.</u> Let $\mathcal{R}(m:n) = \Sigma_i B_n E_i$. Call this the <u>Ree subalgebra</u> of W_n determined by **n**. Let $\mathcal{R}_*(m:n) = \mathcal{R}(m:n) + \operatorname{span}\{E_i^{p^j} \mid (i,j) \in J\}$.

We may grade and filter B_n and W_n by setting

$$\deg x_{ij} = -\deg D_{ij} = p^{j-1} r_i , \quad (i,j) \in I .$$

We say that this filtration (or gradation) is determined by **n** and **r**.

For $\alpha \in A(m:n)$, $(i,j) \in I$, define $\alpha(i,j) \in \mathbb{Z}$, $0 \le \alpha(i,j) < p$, by

$$\alpha(i) = \sum_{j=1}^{n_i} \alpha(i,j) p^{j-1} \quad \text{for all } i .$$

<u>Proposition 1.2.</u> (cf. [13, Theorem 1(b)]) Give $\mathfrak{A}(m:n)$ and $W(m:n)$ the gradations determined

by the m–tuple r and give B_n the gradation determined by n and r. Then:

(a) The linear transformation $\psi:\mathfrak{A}(m:n) \to B_n$ defined by

$$\psi: x^{\alpha} \to \prod_I (x_{ij}^{\alpha(i,j)}/\alpha(i,j)!)$$

is an isomorphism of graded algebras.

(b) The linear transformation Ψ: Der $\mathfrak{A}(m:n) \to W_n$ defined by $\Psi D = \psi D \psi^{-1}$ is an isomorphism of graded algebras.

(c) $\Psi W(m:n) = \mathcal{R}(m:n)$ and $\Psi(\text{Der } W(m:n)) = \mathcal{R}_*(m:n)$.

Proof. It is clear that ψ is an isomorphism of graded vector spaces. Since $\begin{bmatrix}\alpha+\beta\\ \alpha\end{bmatrix} = \prod_i \begin{bmatrix}(\alpha+\beta)(i)\\ \alpha(i)\end{bmatrix}$ by definition and $\begin{bmatrix}(\alpha+\beta)(i)\\ \alpha(i)\end{bmatrix} = \prod_{1\le j\le n_i} \begin{bmatrix}\alpha(i,j)+\beta(i,j)\\ \alpha(i,j)\end{bmatrix}$ by a result of Lucas (cf. [4]), ψ is an isomorphism of algebras. Thus (a) holds. Part (b) is a formal consequence of (a) and (c) holds since $\Psi D_i = E_i$ for all i (as one sees by applying both sides to x_{uv}).

Lemma 1.3. $E_i^{p^k} = \sum_{j=k+1}^{n_i} (-1)^{j+k+1} x_{i,k+1}^{p-1} \cdots x_{i,j-1}^{p-1} D_{ij}$.

Proof. For any $(u,v) \in I$ the images of x_{uv} under the two expressions are the same, proving the equality.

§2. Characterization of $\mathcal{R}_*(m:n)$ by differential forms.

Definition 2.1. For $(a,b) \in J$ set

$$\omega_{ab} = x_{ab}^{p-1}\, dx_{ab} + dx_{a,b+1}.$$

Lemma 2.2. Let $D = \Sigma f_{uv} D_{uv} \in W_n$. then $D\omega_{ab} = d(x_{ab}^{p-1} f_{ab} + f_{a,b+1})$.

Proof. $D\omega_{ab} = -f_{ab} x_{ab}^{p-2} dx_{ab} + x_{ab}^{p-1} df_{ab} + df_{a,b+1} = f_{ab} d(x_{ab}^{p-1}) + x_{ab}^{p-1} df_{ab} + df_{a,b+1}$
$= d(x_{ab}^{p-1} f_{ab} + f_{a,b+1})$.

Proposition 2.3. $\mathcal{R}_*(m:n) = \{D \in W_n \mid D\omega_{ab} = 0 \text{ for all } (a,b) \in J\}$.

Proof. From Definition 1.1 and Lemma 1.3 we see that if $D = f_{uv}D_{uv} \in \mathcal{R}_*(m:n)$ then for any $(a,b) \in J$ we have $x_{ab}^{p-1}f_{ab} + f_{a,b+1} \in F$. Hence, by Lemma 2.2, $\mathcal{R}_*(m:n)\omega_{ab} = (0)$.

Now suppose $D = \Sigma f_{uv}D_{uv}$ satisfies $D\omega_{ab} = 0$ for all $(a,b) \in J$. We must show $D \in \mathcal{R}_*(m:n)$. By subtracting an element of $\mathcal{R}_*(m:n)$ from D we may assume (using Lemma 1.3) that $f_{uv} \in (B_n)_1$ for all $(u,v) \in I$. But then $x_{ab}^{p-1}f_{ab} + f_{a,b+1} \in (B_n)_1$ for all $(a,b) \in J$. Since $d(x_{ab}^{p-1}f_{ab} + f_{a,b+1}) = 0$ by Lemma 2.2 this implies $x_{ab}^{p-1}f_{ab} + f_{a,b+1} = 0$ for all $(a,b) \in J$. This implies $D \in \mathcal{R}(m:n)$, so the lemma is proved.

§3. Automorphisms and the differential forms ω_{ab}.

In this section we prove some technical lemmas involving automorphisms of B_n and the differential forms ω_{ab}. These will be useful in the proof of the main embedding theorem.

Definition 3.1. For $w,z \in B_n$ define $g(w,z) = \sum_{i=1}^{p-1} (-1)^{i+1} w^i z^{p-i}/i$.

Lemma 3.2. (a) If $w,z \in B_n$, then $(w+z)^{p-1}d(w+z) = w^{p-1}dw + z^{p-1}dz + dg(w,z)$.

(b) If $w,z \in (B_n)_1$ then $g(w,z)^2 = 0$.

Proof. For (a) note that $w^{p-1}dw + z^{p-1}dz + dg(w,z) = w^{p-1}dw + z^{p-1}dz +$
$$\sum_{i=1}^{p-1} \{(-1)^{i+1}w^{i-1}z^{p-i}dw + (-1)^i w^i z^{p-1-i}dz\} = \left[\sum_{i=0}^{p-1} (-1)^i w^i z^{p-1-i}\right](dw+dz) = (w+z)^{p-1}d(w+z).$$
If $w,z \in (B_n)_1$, then $w^p = z^p = 0$, so (b) holds.

Now let $(k,\ell) \in J$ and $1 \le i \le m$ be such that $\ell \le n_i$. Let $\mu \in F$. Define

$$y_{kj} = x_{kj} + \mu^{p^{j-1}} x_{i,n_i-\ell+j} \quad \text{for } 1 \le j \le \ell,$$

$$y_{kj} = x_{kj} \quad \text{for } \ell < j \le n_k,$$

$$t_{k1} = 0$$

and

$$t_{k,j+1} = -g(x_{kj},y_{kj}-x_{kj}) - g(y_{kj},t_{kj}) \text{ for } 0 < j < n_k .$$

Let θ be the automorphism of B_n defined by

$$\theta x_{kj} = y_{kj} + t_{kj} \text{ for } 1 \le j \le n_k ,$$

and

$$\theta x_{uv} = x_{uv} \text{ for } (u,v) \in I,\ u \ne k .$$

Lemma 3.3. Let (k,ℓ), i and θ be as above. Then

$$\theta\omega_{kj} = \omega_{kj} + \mu^{p^j}\omega_{i,n_i-\ell+j} \text{ for } 1 \le j < \ell ,$$

$$\theta\omega_{k\ell} = \omega_{k\ell} + \mu^{p^\ell} x_{i,n_i}^{p-1} dx_{i,n_i} ,$$

$$\theta\omega_{kj} = \omega_{kj} \text{ for } \ell < j < n_k ,$$

and

$$\theta\omega_{uv} = \omega_{uv} \text{ for } (u,v) \in J,\ u \ne k .$$

Proof. Observe that $t_{kj}^{p-1}dt_{kj} = 0$ (by Lemma 3.2(b) if $p > 3$ and by direct computation if $p = 2,3$). Then, as the last assertion of the lemma is obvious, it is sufficient to use Lemma 3.2 to compute:

$$\theta\omega_{kj} = (y_{kj}+t_{kj})^{p-1}d(y_{kj}+t_{kj}) + d(y_{k,j+1}+t_{k,j+1})$$

$$= y_{kj}^{p-1}dy_{kj} + dg(y_{kj},t_{kj}) + dy_{k,j+1} + dt_{k,j+1}$$

$$= (x_{kj} + (y_{kj}-x_{kj}))^{p-1}d(x_{kj}+(y_{kj}-x_{kj})) + dx_{k,j+1}$$

$$+ d(y_{k,j+1}-x_{k,j+1}) - dg(x_{kj},y_{kj}-x_{kj})$$

$$= x_{kj}^{p-1}dx_{kj} + dx_{k,j+1} + (y_{kj}-x_{kj})^{p-1}d(y_{kj}-x_{kj}) + d(y_{k,j+1}-x_{k,j+1})$$

$$= \omega_{kj} + (y_{kj}-x_{kj})^{p-1}d(y_{kj}-x_{kj}) + d(y_{k,j+1}-x_{k,j+1}) .$$

It is immediate from the definition of the y_{kj} that this has the required value.

Note that if B_n is given the gradation determined by **n** and **r** then Ω_n has corresponding filtration and gradation and $\omega_{ab} \in (\Omega_n)_{p^b r_a}$.

Lemma 3.4. Give B_n and W_n the filtrations and gradations determined by the m–tuples **n** and **r**. Suppose that there exist an integer $s > 0$, elements $G_u \in W_n$, $1 \le u \le m$, and elements $g^u_{ab} \in (B_n)_{[p^b r_a - r_u + s]}$ for each $1 \le u \le m$, $(a,b) \in J$, such that

$$G_u \equiv E_u \mod (W_n)_{-r_u+1},$$

$$G_u \omega_{ab} \equiv dg^u_{ab} \mod (\Omega_n)_{p^b r_a - r_u + s+1}$$

and

$$E_u g^v_{ab} = E_v g^u_{ab}$$

for all $1 \le u, v \le m$, $(a,b) \in J$. Then there exists an automorphism $\Phi \in Aut_1(W_n)$ (i.e., such that $(\Phi - I)(W_n)_t \subseteq (W_n)_{t+1}$ for all t) such that

$$(\Phi G_u)\omega_{ab} \in (\Omega_n)_{p^b r_a - r_u + s+1}$$

for all $1 \le u \le m$, $(a,b) \in J$.

Proof. It is easily seen (using the identification of B_n with $\mathfrak{A}(m:n)$ of Proposition 1.2) that there exist elements $h_{ab} \in (B_n)_{[p^b r_a + s]}$ for all $(a,b) \in J$ such that $g^u_{ab} + E_u h_{ab} \in F x^{p-1}_{u1} \cdots x^{p-1}_{u,n_u}$ for all u. Then we define elements $f_{ab} \in B_n$ for $(a,b) \in I$ by

$$f_{a1} = 0$$

and

$$f_{a,b+1} = -x^{p-1}_{ab} f_{ab} - h_{ab}$$

for $(a,b) \in J$. Writing $D = \Sigma f_{ab} D_{ab}$ we have $E_u D\omega_{ab} = E_u d(x^{p-1}_{ab} f_{ab} + f_{a,b+1}) = -E_u d(h_{ab}) = -d(E_u h_{ab})$ for $1 \le u \le m$, $(a,b) \in J$. Clearly $D \in (W_n)_{[s]}$. Now let Γ be an automorphism of W_n satisfying $(\Gamma - I - \mathrm{ad}\, D)((W_n)_t) \subseteq (W_n)_{t+s+1}$ for all t. (Such an automorphism exists by [14].) Then, modulo $(\Omega_n)_{p^b r_a - r_u + s+1}$ we have $(\Gamma G_u)\omega_{ab} \equiv (G_u + [D,G_u])\omega_{ab} \equiv (G_u + [D,E_u])\omega_{ab} \equiv G_u\omega_{ab}$ $- E_u D\omega_{ab} \equiv dg^u_{ab} + d(E_u h_{ab})$. By the choice of h_{ab} this is an element of $Fd(x^{p-1}_{u1} \cdots x^{p-1}_{u,n_u})$.

Therefore, replacing G_u by ΓG_u, we may assume $g^u_{ab} = \lambda^u_{ab} x^{p-1}_{u1} \cdots x^{p-1}_{u,n_u}$ for some $\lambda^u_{ab} \in F$. Note that, because $g^u_{ab} \in (B_n)_{[p^b r_a - r_u + s]}$, if $\lambda^u_{ab} \neq 0$ then we have $(p^{n_u}-1)r_u = p^b r_a - r_u + s$ so $p^b r_a < p^{n_u} r_u$.

Now suppose that $\lambda^i_{k\ell} \neq 0$. Take $\mu \in F$ so that $\mu^{p^\ell} = (-1)^{n_i} \lambda^i_{k\ell}$ and let θ be the automorphism of Lemma 3.3. Let Θ be the automorphism of W_n defined by $\Theta D = \theta D \theta^{-1}$ for all $D \in W_n$. Since $p^\ell r_k < p^{n_i} r_i$ we have $\theta \in \mathrm{Aut}_1(B_n)$ and so $\Theta, \Theta^{-1} \in \mathrm{Aut}_1(W_n)$. It is clear from Lemma 3.3 that

$$(\Theta^{-1} G_u)\omega_{ab} \equiv G_u\omega_{ab} \mod (\Omega_u)_{p^b r_a - r_u + s+1}$$

for all $1 \le u \le m$, $(a,b) \in J$ except $u = i$, $(a,b) = (k,\ell)$. (Unless $a = k$, $1 \le b < \ell$ this is obvious. In that case note that $G_u\omega_{i,n_i-\ell+b} \in (\Omega_n)_{p^{n_i-\ell+b} r_i - r_u + s}$. Since $\lambda^i_{k\ell} \neq 0$ we have $p^{n_i} r_i > p^\ell r_k$ and so $p^{n_i-\ell+b} r_i - r_u + s > p^b r_k - r_u + s$ giving the required result.) Also $(\Theta^{-1} G_i)\omega_{k\ell} = \theta^{-1} G_i \theta\omega_{k\ell} =$ $\theta^{-1} G_i(\omega_{k\ell} + (-1)^{n_i} \lambda^i_{k\ell} x^{p-1}_{i,n_i} dx_{i,n_i}) \equiv \theta^{-1}(dg^i_{k\ell} - \lambda^i_{k\ell} dx^{p-1}_{i1} \cdots x^{p-1}_{i,n_i}) = 0$, where the congruence is modulo $(\Omega_n)_{p^\ell r_k - r_i + s+1}$. Thus we may change $\lambda^i_{k\ell}$ to 0 without changing the other λ^u_{ab}. Thus we may assume all $\lambda^u_{ab} = 0$, so the lemma is proved.

<u>§4. Conjugation of subalgebras of</u> W_n <u>into</u> $\mathscr{R}(m{:}n)$.

We now prove a conjugacy result for subalgebras of W_n.

<u>Proposition 4.1.</u> Let $r = (r_1, \cdots, r_m)$ where $1 \le r_i \le (p-1)/2$ for $1 \le i \le m$. Give W_n the filtration determined by the m-tuples n and r. Let $L \subseteq W_n$ be a subalgebra such that

(*) $\{E_1, \cdots, E_m\}$ is a basis for $\sum_{i<0} (\mathrm{gr}\, L)_i$.

Then there exists an automorphism Φ of the filtered algebra W_n such that $\Phi L \subseteq \mathcal{R}(m:n)$.

Proof. For each integer $s \geq 0$ let $P(s)$ denote the condition that (*) is satisfied and

$$L_i\,\omega_{ab} \subseteq (\Omega_n)_{p^b r_a+i+s}$$

for all $i \in Z$, $(a,b) \in J$. Also let $Q(s)$ denote the condition that $P(s)$ holds and there exist elements $G_i \in L$, $1 \leq i \leq m$, such that

$$G_i \equiv E_i \mod (W_n)_{-r_i+1}$$

and

$$G_i\omega_{ab} \in (\Omega_n)_{p^b r_a - r_i + s+1}$$

for $1 \leq i \leq m$. Note that $Q(0)$ holds by hypothesis.

We claim that $Q(s)$ implies $P(s+1)$. To see this suppose that for some integer t we have $L_i\omega_{ab} \subseteq (\Omega_n)_{p^b r_a+i+s+1}$ for all $i < t$. (Since $Q(s)$ holds this is true for $t = 0$.) Let $D = \Sigma f_{uv}D_{uv} \in L_t$ and $g_{ab} = x_{ab}^{p-1}f_{ab} + f_{a,b+1}$ for $(a,b) \in J$. Write $q = p^b r_a+s$. Then for $1 \leq i \leq m$ by assumption we have $[G_i,D]\omega_{ab} \in (\Omega_n)_{q+t+1-r_i}$ and by $Q(s)$ we have $DG_i\omega_{ab} \in (\Omega_n)_{q+t+1-r_i}$. Hence $G_iD\omega_{ab} \in (\Omega_n)_{q+t+1-r_i}$ for all i, $1 \leq i \leq m$. As $E_i \equiv G_i$ $\mathrm{mod}(W_n)_{-r_i+1}$ this implies $E_iD\omega_{ab} \in (\Omega_n)_{q+t+1-r_i}$. Thus $E_ig_{ab} \in (B_n)_{q+t+1-r_i} + F$. However, we also have $E_ig_{ab} \in (B_n)_{p^b r_a - r_i+t}$. Since $b \geq 1$ and $r_i < p$ this gives $E_ig_{ab} \in (B_n)_1$, so $E_ig_{ab} \in (B_n)_{q+t+1-r_i}$. Therefore $g_{ab} \in (B_n)_{q+t+1} + F$. Since $g_{ab} \in (B_n)_{p^b r_a+t} \subseteq (B_n)_1$ this implies $g_{ab} \in (B_n)_{q+t+1}$. Thus $D\omega_{ab} = dg_{ab} \in (\Omega_n)_{q+t+1} = (\Omega_n)_{p^b r_a+t+s+1}$. By induction on t, $P(s+1)$ holds.

We now show that if $P(s)$ holds for L then there exists $\Phi \in \mathrm{Aut}_1(W_n)$ such that $Q(s)$ holds for ΦL. We continue to write $q = p^b r_a+s$. For $1 \leq u \leq m$, $(a,b) \in J$, define $g_{ab}^u \in (B_n)_{[q-r_i]}$ by

$$G_u\omega_{ab} \equiv dg_{ab}^u \ \mathrm{mod}(\Omega_n)_{[q-r_u+1]}\,.$$

Then it is clear that

$$[G_u,G_v]\omega_{ab} \equiv d(E_u g^v_{ab} - E_v g^u_{ab}) \bmod (\Omega_n)_{[q-r_u-r_v+1]}\,.$$

However, as $[E_u,E_v] = 0$ we have $[G_u,G_v] \in L_{-r_u-r_v+1}$ and so by P(s) we have $[G_u,G_v]\omega_{ab} \in (\Omega_n)_{[q-r_u-r_v+1]}$. Thus $E_u g^v_{ab} - E_v g^u_{ab} \in F$ for all $1 \le u, v \le m$, $(a,b) \in J$. But $E_u g^v_{ab} - E_v g^u_{ab} \in (B_n)_{[q-r_u-r_v]}$. Since $r_u, r_v \le (p-1)/2$ we have $q-r_u-r_v = p^b r_a + s - r_u - r_v \ge 1$ so $E_u g^v_{ab} - E_v g^u_{ab} \in (B_n)_1$. Thus $E_u g^v_{ab} = E_v g^v_{ab}$ for all $1 \le u, v \le m$, $(a,b) \in J$. Then Lemma 3.4 gives the required Φ.

Now let $t \in \mathbb{Z}$ be so large that $(\Omega_n)_t = (0)$. By induction there exists $\Phi \in \mathrm{Aut}_1(W_n)$ such that P(t) holds for ΦL. Since $(\Omega_n)_t = 0$ this gives (by Proposition 2.3) $\Phi L \subseteq \mathcal{R}_*(m:n)$. As $\{E_1,\dots,E_m\}$ is a basis for $\sum_{i<0} (\mathrm{gr}\, L)_i$ and $\deg(E_u^{p^v}) < \deg E_w$ for any $1 \le u, w \le m$, $v \ge 1$ (as all $r_i < p$) we see that $\Phi L \subseteq \mathcal{R}(m:n)$, as required.

§5. The W(m:n) embedding theorem.

In this section we will prove (Theorem 5.3) that a subalgebra $L \subseteq W(m)$ satisfying certain hypotheses can be conjugated into $W(m:n)$. In view of Proposition 4.1 it will be sufficient to find an appropriate embedding of L into W_n. We do this by embedding L in a restricted Lie algebra M with a restricted subalgebra M_0 of codimension n and by using Blattner's technique [1] to embed M into $\mathrm{Der\,Hom}_{uM_0}(uM,F) \cong W_n$. That Blattner's technique works for restricted Lie algebras (using the restricted enveloping algebra uM in place of the universal enveloping algebra as in Blattner's work) is an observation of Kac [6]. The argument of [2] shows that the embedding is filtration preserving.

We begin by describing the Blattner–Kac technique. Let M be a restricted Lie algebra with a restricted filtration (i.e., if $x \in M_i$ then $x^p \in M_{pi}$). Let $s = (s_1,\dots,s_n)$ be an n–tuple of positive integers and $\{m_1,\dots,m_n\} \subseteq M$ be such that for every $t \ge 1$ $\{m_i + M_0 \mid s_i \le t\}$ is a basis for M_{-t}/M_0. Give uM the usual coalgebra structure. Filter uM by $(uM)_0 = u(M_0)$, $(uM)_\ell = (0)$ and $(uM)_{-\ell} = \mathrm{span}\{t m_{i_1} \cdots m_{i_q} \mid t \in (uM)_0,\ q \ge 0,\ s_{i_1} + \cdots + s_{i_q} \le \ell\}$ for $\ell > 0$. Then the algebra $V = \mathrm{Hom}_{uM_0}(uM,F)$ (which is isomorphic to B_n) is filtered by setting

$$V_\ell = \{\varphi \in V \mid \varphi((uM)_{-\ell+1}) = (0)\}\,.$$

For an n–tuple β of integers with $0 \le \beta(j) \le p-1$ for all j define $m^\beta = m_1^{\beta(1)} \cdots m_n^{\beta(n)}$. For $m \in M$, $\varphi \in V$ define

$$\gamma(m)\colon V \to V$$

by

$$(\gamma(m)\varphi)(v) = \varphi(vm)$$

for $\varphi \in V$, $v \in uM$.

Theorem 5.1. (Blattner–Kac). Let M and V be as above. For $m \in M$, $\gamma(m) \in \mathrm{Der}\, V$ and $\gamma\colon M \to \mathrm{Der}\, V$ is a homomorphism of filtered Lie algebras. Also, $\ker \gamma$ is the largest ideal of M contained in M_0. If $M_{i+1} = \{x \in M_i \mid [x,M_{-1}] \subseteq M_i\}$ for all $i \ge 0$ and $M_t = (0)$ for some t then γ is an injection of filtered Lie algebras.

As noted above this result is obtained by replacing the use of the universal enveloping algebra in Blattner's result [1, Theorem 1] (and in [2] for the statements about the filtration) by uM.

Proposition 5.2. Let M be a filtered restricted Lie algebra. Let L be a filtered subalgebra such that L_0 contains no nonzero ideals of L. Suppose that there exist $\ell_1,\dots,\ell_m \in L$ and m–tuples of positive integers $\mathbf{n}$ and $\mathbf{r}$ such that $r_i \le (p-1)/2$ for $1 \le i \le m$ and that, for every $t \ge 1$, $\{\ell_i + L_0 \mid i \le i \le m, r_i \le t\}$ is a basis for L_{-t}/L_0 and $\{\ell_i^{p^j} + M_0 \mid 1 \le i \le m, 0 \le j < n_i, p^j r_i \le t\}$ is a basis for M_{-t}/M_0. Let $n = n_1 + \cdots + n_m$. Suppose also that $[\ell_i,\ell_j] \subseteq L_{-r_i-r_j+1}$,

$\ell_i^{p^{n_i}} \in M_{-p^{n_i} r_i+1}$ for $1 \le i, j \le m$ and $L_{t+1} = \{x \in L_t \mid [x,L_{-1}] \subseteq L_t\}$ for $t \ge 0$. Give W_n the filtration determined by $\mathbf{n}$ and $\mathbf{r}$. Then there is an injection of filtered Lie algebras $\lambda\colon L \to W_n$ such that $\lambda(L) \subseteq \mathscr{R}(m:\mathbf{n})$.

Proof. In view of Proposition 4.1 it is sufficient to find an injection of filtered Lie algebras $\nu\colon L \to W_n$ such that $\nu(\ell_i) \equiv E_i \bmod (W_n)_{-r_i+1}$ for $1 \le i \le m$. If we write $\ell^\beta = \ell_1^{\beta(1)} \cdots \ell_m^{\beta(m)}$

we see (by the assumptions on $[\ell_i,\ell_j]$ and $\ell_i^{p^{n_i}}$) that $\ell^\beta \ell_i \equiv \ell^{\beta+\varepsilon_i} \bmod (uM)_{-\Sigma\beta(j)r_j-r_i+1}$ if $\beta(i) < p^{n_i}-1$ and $\ell^\beta \ell_i \equiv 0 \bmod (uL)_{-\Sigma\beta(j)r_j-r_i+1}$ if $\beta(i) = p^{n_i}-1$. Then if we define $\pi\colon B_n \to \mathrm{Hom}_{uM_0}(uM,F)$ by $\pi(\prod x_{uv}^{\alpha(u,v)}/\alpha(u,v)!)(\ell^\beta) = \delta_{\alpha\beta}$ (where the $\alpha(u,v)$ are as in §1) and let $v(\ell) = \pi^{-1}\gamma(\ell)\pi$, it is immediate that $v(\ell_i) \equiv E_i \bmod (W_n)_{-r_i+1}$, as required.

Theorem 5.3. Let $r = (r_1,\ldots,r_m)$ be an m-tuple of integers with $1 \le r_i \le (p-1)/2$ for all i. Give $W(m)$ the filtration determined by r. Let $L \subseteq W(m)$ satisfy $L_0 = L \cap \mathfrak{A}(m)_1 W(m)$, $\dim L/L_0 = m$ and center $L = (0)$. Assume that gr L is a transitive subalgebra of $W(m:\mathbf{n})$ and that every derivation of gr L extends to a derivation of $W(m:\mathbf{n})$. Then there is an automorphism Φ of the filtered Lie algebra $W(m)$ such that $\Phi L \subseteq W(m)$.

Proof. Since center $L = (0)$ we have $L \subseteq \mathrm{Der}\, L$. Let $\mathcal{L}$ be the restricted subalgebra of Der L generated by L. Now Der L is filtered by $(\mathrm{Der}\, L)_i = \{D \in \mathrm{Der}\, L \mid DL_j \subseteq L_{i+j}$ for all $j\}$. Since gr L is transitive we have $(\mathrm{Der}\, L)_i \cap L = L_i$.

Let $\ell_i \in L_{-r_i}$ satisfy $\ell_i \equiv E_i \bmod W(m)_{-r_i+1}$. We claim that the hypotheses of Proposition 5.2 are satisfied where $M = \mathcal{L}$. To see this it is only necessary to show that $\{\ell_i^{p^j} + (\mathcal{L})_0 \mid 1 \le i \le m, 0 \le j < n_i, p^j r_i \le t\}$ spans M_{-t}/M_0 for every $t \ge 1$. Denote the span of this set by N_{-t}/M_0 and suppose that $N_{-t}/M_0 = M_{-t}/M_0$ for all $t \le s$. (This holds for $s = 0$.) Let $D \in M_{-t-1}$. Then D induces a derivation of gr L. If this derivation has degree $\ge -t$ then $D \in M_{-t}$ (by definition of the filtration on Der L). So $D+M_0 \in N_{-t}/M_0 \subseteq N_{-t-1}/M_0$. If the derivation induced by D has degree $-t-1$ then its extension to $W(m:\mathbf{n})$ must be a linear combination of derivations $(\mathrm{ad}\, D_i)^{p^j}$ with $p^j r_i = t+1$. Subtracting the corresponding linear combination of $\ell_i^{p^j}$ from D we obtain an element $D' \in M_{-t}$. Thus $D+M_0 \in N_{-t-1}/M_0$. Thus $M_{-t-1}/M_0 = N_{-t-1}/M_0$ and so $M_{-t}/M_0 = N_{-t}/M_0$ for all t by induction.

Hence the hypotheses of Proposition 5.2 are satisfied. That proposition then shows that L is isomorphic (as a filtered Lie algebra) to a subalgebra of $W(m:\mathbf{n})$. Using Theorem 2.2 of [2] we obtain $\Phi L \subseteq W(m:\mathbf{n})$ for some $\Phi \subseteq \mathrm{Aut}\, W(m)$, proving the theorem.

We remark that if, for X = W,S,CS,H,CH or K we have $X(m:\mathbf{n})^{(2)} \subseteq \mathrm{gr}\, L \subseteq X(m:\mathbf{n})$ then the hypotheses of Theorem 5.3 are clearly satisfied.

References.

[1] R. Blattner, Induced and produced representations of Lie algebras, Trans. Amer. Math. Soc. 144 (1969), 457–474.

[2] R. E. Block and R. L. Wilson, On filtered Lie algebras and divided power algebras, Comm. Algebra 3 (1975), 571–589.

[3] ______________________________, Classification of the restricted simple Lie algebras, J. Algebra 114 (1988), 115–259.

[4] N. J. Fine, Binomial coefficients modulo a prime, Amer. Math. Monthly 54 (1947), 589–592.

[5] V. G. Kac, Global Cartan pseudogroups and simple Lie algebras of characteristic p [Russian], Uspehi Mat. Nauk 26 (1971), 199–200.

[6] __________, Description of filtered Lie algebras with which graded Lie algebras of Cartan type are associated, Izv. Akad. Nauk. SSSR Ser. Mat. 38 (1974), 800–834; Errata, 40 (1976), 1415 [Russian]; Math. USSR–Izv. 8 (1974), 801–835; Errata, 10 (1976), 1339 [English transl.].

[7] A. I. Kostrikin and I. R. Šafarevič, Graded Lie algebras of finite characteristic, Izv. Akad. Nauk. SSSR Ser. Mat. 33 (1969), 251–322 [Russian]; Math. USSR–Izv. 3 (1969), 237–304 [English transl.].

[8] D. E. Radford, Divided power structures on Hopf algebras and embedding Lie algebras into special–derivation algebras, J. Algebra 98 (1986), 143–170.

[9] R. Ree, On generalized Witt algebras, Trans. Amer. Math. Soc. 83 (1956), 510–546.

[10] S. Serconek and R. L. Wilson, Classification of forms of restricted simple Lie algebras of Cartan type, Comm. Algebra, to appear.

[11] H. Strade and R. Farnsteiner, "Modular Lie Algebras and Their Representations," Marcel Dekker, Vol. 116, New York, 1988.

[12] R. L. Wilson, Nonclassical simple Lie algebras, Bull. Amer. Math. Soc. 75 (1969), 987–991.

[13] _____________, Classification of generalized Witt algebras over algebraically closed fields, Trans. Amer. Math. Soc. 153 (1971), 191–210.

[14] _____________, Automorphisms of graded Lie algebras of Cartan type, Comm. Algebra 3 (1975), 591–613.

[15] _____________, A structural characterization of the simple Lie algebras of generalized Cartan type over fields of prime characteristic, J. Algebra 40 (1976), 418–465.

1980 AMS subject classification (1985 revision) 17B50, 17B20

School of Mathematics
Institute of Advanced Study
Princeton, NJ 08540, USA

Department of Mathematics
Rutgers University
New Brunswick, NJ 08903, USA

This paper is in final form and no version of it will appear elsewhere.

ISOMORPHISM CLASSES OF HAMILTONIAN LIE ALGEBRAS

G. M. Benkart, T. B. Gregory, J. M. Osborn, H. Strade, and R. L. Wilson (*)

Over an algebraically closed field F of prime characteristic $p \geq 7$ the known finite-dimensional simple Lie algebras are classical (i.e. analogues of the finite-dimensional simple complex Lie algebras) or are members of one of the four series of Lie algebras of Cartan type: the general, special, Hamiltonian, and contact series. The Lie algebras in the first two Cartan series have been determined up to isomorphism [T], [W, 4]. (Note that the results in the latter two papers give slightly different isomorphism classes for the Lie algebras in the special series.) The main result of this paper (Theorem 1) deals with the problem of determining the isomorphism classes of the Lie algebras of the Hamiltonian series. Unless noted otherwise, our notation is that of [W,3]. Throughout the paper we let m be a positive integer and $\ell = [m/2]$. For $1 \leq i \leq \ell$ we write $\tilde{i} = i+\ell$, $\widetilde{i+\ell} = i$, and $\pi(i) = -\pi(\tilde{i}) = 1$. For $1 \leq i \leq m$, we let ε_j denote the m-tuple $(\delta_{1j}, \ldots, \delta_{mj})$ where δ_{ij} is the Kronecker delta.

<u>Theorem 1.</u> Let H be a Lie algebra of the Hamiltonian series of Lie algebras of Cartan type over an algebraically closed field F of characteristic $p > 3$. Then there exist an even integer $m = 2\ell$ and an m-tuple of positive integers $\underline{n}$ such that H is isomorphic to the Lie algebra of all elements of $W(m:\underline{n})$ annihilating the

---

(*) This paper was written during the Special Year of Lie Algebras at the University of Wisconsin, Madison. The authors would like to thank the National Science Foundation for its support of this special year through grant #DMS-87-02928. In addition, RLW acknowledges partial support from The Institute for Advanced Study and from NSF grant #DMS-86-03151 and GMB and JMO from NSF grant #DMS-85-01514.

This paper is in final form and no version of it will appear elsewhere.

skew-symmetric differential 2-form

$$\omega = \sum_{1 \leq i,j \leq m} \omega_{ij}\, dx_i \wedge dx_j,$$

where ω has one of two possible forms. Either

(a) $\omega_{ij} \in \mathcal{O}(m:\underline{n})$ for all i and j and

$$\omega_{ij} = \pi(i)\delta_{j\tilde{i}} + a_{ij}\, x^{(p^{n_i}-1)\varepsilon_i} x^{(p^{n_j}-1)\varepsilon_j}$$

where (a_{ij}) is a skew-symmetric $m \times m$ matrix over F, or

(b) $\omega_{ij} \notin \mathcal{O}(m:\underline{n})$ for some i and j, in which case there exists an integer s with $1 \leq s \leq m$ such that

$$\omega_{ij} = \exp(x^{p^{n_s}\varepsilon_s}) \Big(\pi(i)\delta_{i\tilde{j}} + (\delta_{is}-\delta_{js})^2 \big(a_{ij}\, x^{(p^{n_i}-1)\varepsilon_i} x^{(p^{n_j}-1)\varepsilon_j}\big)$$

$$+ \tfrac{1}{2}\delta_{js}\pi(i) x^{(p^{n_s}-1)\varepsilon_s} x^{\varepsilon_{\tilde{i}}} - \tfrac{1}{2}\delta_{is}\pi(j) x^{(p^{n_s}-1)\varepsilon_s} x^{\varepsilon_{\tilde{j}}} \Big),$$

where (a_{ij}) is a skew-symmetric $m \times m$ matrix over F with $a_{s\tilde{s}} = 0 = a_{\tilde{s}s}$.

<u>Remark.</u> The case when $\omega_{ij} \in \mathcal{O}(m:\underline{1})$ for all i,j has been investigated in [K] and more recently in [K,K].

The proof of this theorem is based on Kac's general result concerning isomorphisms between algebras of Cartan type. Before stating this result we recall some notation involving the divided power algebra $\mathcal{O}(m)$. The algebra $\mathcal{O}(m)$ is filtered by setting $\deg x_i = 1$ for all i so that $\mathcal{O}(m) = \mathcal{O}(m)_0 \supset \mathcal{O}(m)_1 \supset \dots$. For any $u \in \mathcal{O}(m)_1$, the exponential of u is defined by $\exp(u) = \Sigma_{j \geq 0}\, u^{(j)}$ where $u^{(j)}$ is the image of u under the jth divided power map. Let $\mathrm{Aut}_{\mathrm{div}}\mathcal{O}(m)$ denote the group of automorphisms of $\mathcal{O}(m)$ <u>as a divided power algebra</u>. Thus, $\mathrm{Aut}_{\mathrm{div}}\mathcal{O}(m)$ is the group of all algebra automorphisms φ of $\mathcal{O}(m)$ such that $\varphi(u^{(j)}) = (\varphi u)^{(j)}$ for all $u \in \mathcal{O}(m)_1$ and all $j \geq 0$. Equivalently, it is the set of all algebra automorphisms φ of

$\mathfrak{A}(m)$ such that $\varphi W(m)\varphi^{-1} \subseteq W(m)$. This group is denoted by $\mathrm{Aut}(\mathfrak{A}(m){:}W(m))$ in [W,3].

A simple Lie algebra of Cartan type X ($X = W, S, H$, or K) is determined by a triple $(m, \underline{\underline{n}}, \omega)$ where m is an integer ≥ 1, $\underline{\underline{n}} = (n_1, \ldots, n_m)$ is an m-tuple of integers ≥ 1, and ω is a differential form on the divided power algebra $\mathfrak{A}(m)$ which is conjugate under $\mathrm{Aut}_{div}\mathfrak{A}(m)$ to the standard differential form ω_X given by

$$\omega_W = 0,$$

$$\omega_S = dx_1 \wedge \ldots \wedge dx_m$$

$$\omega_H = \sum_{i=1}^{\ell} dx_i \wedge dx_{\tilde{i}}, \qquad m = 2\ell \text{ even,}$$

$$\omega_K = dx_m + \sum_{i=1}^{2\ell} \pi(i)\, x_i\, dx_{\tilde{i}}, \qquad m = 2\ell+1 \text{ odd,}$$

where $\tilde{i}$ and $\pi(i)$ are as in Theorem 1 for ω_H and ω_K. Then $X(m{:}\underline{\underline{n}}{:}\omega) = \{ D \in W(m{:}\underline{\underline{n}}) \mid D\omega = 0 \}$ for $X = W, S, H$ and $K(m{:}\underline{\underline{n}}{:}\omega) = \{D \in W(m{:}\underline{\underline{n}}) \mid D\omega \in \mathfrak{A}(m{:}\underline{\underline{n}})\omega\}$. Furthermore the compatibility condition

$$X(m{:}\underline{\underline{n}})^{(2)} \subseteq gr\, X(m{:}\underline{\underline{n}}{:}\omega) \subseteq X(m{:}\underline{\underline{n}})$$

must be satisfied.

If $\sigma \in \mathcal{S}ym\{1, \ldots, m\}$ then there is an automorphism of $\mathfrak{A}(m)$, again denoted by σ, defined by $\sigma(x_i) = x_{\sigma(i)}$ for all i. If $\underline{\underline{n}} = (n_1, \ldots, n_m)$, we define $\underline{\underline{n}}_\sigma = (n_{\sigma(1)}, \ldots, n_{\sigma(m)})$. Then $\sigma\mathfrak{A}(m{:}\underline{\underline{n}}) = \mathfrak{A}(m{:}\underline{\underline{n}}_{\sigma^{-1}})$. We define an equivalence relation $\sim$ on the set of triples $(m, \underline{\underline{n}}, \omega)$ which satisfy the compatibility condition by saying $(m, \underline{\underline{n}}, \omega) \sim (m', \underline{\underline{n}}', \omega')$ if and only if $m = m'$ and there exist $\sigma \in \mathcal{S}ym\{1, \ldots, m\}$, $\varphi \in \{\psi \in \mathrm{Aut}_{div}\mathfrak{A}(m) \mid \psi(\mathfrak{A}(m, \underline{\underline{n}})) \subseteq \mathfrak{A}(m, \underline{\underline{n}})\}$ and $0 \neq c \in F$ such that $\underline{\underline{n}}' = \underline{\underline{n}}_{\sigma^{-1}}$ and $\sigma\varphi\omega = c\omega'$. Then Kac's theorem on isomorphisms states that the algebras determined by $(m, \underline{\underline{n}}, \omega)$ and $(m', \underline{\underline{n}}', \omega')$ are isomorphic if and only if $(m, \underline{\underline{n}}, \omega) \sim (m', \underline{\underline{n}}', \omega')$.

In the case of type H algebras we must have m even and ω a skew-symmetric 2-form. Thus, $\omega = \Sigma_{i,j}\, \omega_{ij}\, dx_i \wedge dx_j$ where $\omega_{ji} = -\omega_{ij}$ for all $i,j = 1, \ldots, m$. Then ω is determined by the skew-symmetric matrix $M(\omega) = (\omega_{ij})$ with entries in $\mathfrak{A}(m)$. The requirement that $\omega = \varphi\omega_H$ for some $\varphi \in \mathrm{Aut}_{div}\mathfrak{A}(m)$ is equivalent, according to [K, Lemma 2.0(2)], to having the following hold:

$$(1)\quad \begin{array}{l} M(\omega) \text{ is skew-symmetric and nonsingular} \\ D_i\omega_{jk} + D_j\omega_{ki} + D_k\omega_{ij} = 0 \qquad \text{for all } i,j,k. \end{array}$$

We refer to (1) as the Hamiltonian condition. Lemma 2.0(4) of [K] shows that the compatibility condition may be expressed in terms of $M(\omega)$ as

$$(2)\qquad M(\omega) = f\,B$$

where $f \in \mathcal{O}(m)$ is invertible, $B \in \mathfrak{Mat}_{m\times m}(\mathcal{O}(m:\underline{n}))$, and $f^{-1}D_i f \in \mathcal{O}(m:\underline{n})$ for all i.

Thus, our object of study will be the set of triples $\mathcal{T}_H = \{(m, \underline{n}, \omega) \mid (1)$ and (2) hold for $M(\omega)\}$ with the equivalence relation $\sim$. A complete solution of the isomorphism problem for the Hamiltonian algebras (which we do not obtain here) would consist of determining the $\sim$ equivalence classes of $\mathcal{T}_H$. In this paper we find a reasonably small subset $\mathcal{S}_H \subseteq \mathcal{T}_H$ such that every $\sim$ equivalence class in $\mathcal{T}_H$ meets $\mathcal{S}_H$. However, distinct elements in $\mathcal{S}_H$ may be $\sim$ equivalent, and we do not determine the $\sim$ equivalence classes in $\mathcal{S}_H$.

To describe $\mathcal{S}_H$ we introduce some notation. Let $J_0 \in \mathfrak{Mat}_{m\times m}(F)$ be defined by

$$J_0 = (\pi(i)\,\delta_{i\tilde{j}}) = \begin{bmatrix} 0 & I \\ -I & 0 \end{bmatrix}.$$

Let γ_i and $\gamma_{ij} \in A(m:\underline{n}) = \{\alpha = (\alpha_1,\ldots,\alpha_n) \mid 0 \le \alpha_i < p^{n_i}\}$ be given by $\gamma_i = (p^{n_i}-1)\varepsilon_i$ and $\gamma_{ij} = \gamma_i + \gamma_j$. For $1 \le s \le m$ define $U_s \in \mathfrak{Mat}_{m\times m}(\mathcal{O}(m:\underline{n}))$ by $(U_s)_{ij} = \delta_{is}x^{\gamma_s}$, and $V \in \mathfrak{Mat}_{m\times m}(\mathcal{O}(m:\underline{n}))$ by $V_{ij} = \frac{1}{2}\,\delta_{ij}\,\pi(i)\,x^{\varepsilon_{\tilde{i}}}$. Let $J_s = J_0 + VU_s^t - U_sV$. Set $e_s = 1$ when $s = 0$, and $e_s = \exp(x^{p^{n_s}\varepsilon_s})$ for $1 \le s \le m$. We claim that $e_s J_s$ satisfies the Hamiltonian condition. Indeed, e_sJ_s is skew-symmetric and nonsingular. If the indices are not distinct, or if none of the indices i, j, or k is s, then it is immediate that $D_i(e_s\xi_{jk}) + D_j(e_s\xi_{ki}) + D_k(e_s\xi_{ij}) = 0$ where $\xi_{ij} = (J_s)_{ij}$. Similarly, if $k = s$ and $j \ne \tilde{i}$, we again get 0. Finally,

$D_i(e_s\xi_{\tilde{i}s}) + D_{\tilde{i}}(e_s\xi_{si}) + D_s(e_s\xi_{i\tilde{i}})$

$$= \tfrac{1}{2} e_s \pi(\tilde{i}) x^{(p^{n_s}-1)\varepsilon_s} - \tfrac{1}{2} e_s \pi(i) x^{(p^{n_s}-1)\varepsilon_s} + \pi(i) e_s x^{(p^{n_s}-1)\varepsilon_s} = 0$$

to give the assertion.

If $A = (a_{ij}) \in \mathfrak{Mat}_{m\times m}(F)$, define $A^{\underline{n}} \in \mathfrak{Mat}_{m\times m}(\mathfrak{Q}(m:\underline{n}))$ by $A^{\underline{n}} = \Gamma^{\underline{n}} A \Gamma^{\underline{n}}$ where $\Gamma^{\underline{n}} = \mathrm{diag}(x^{\gamma_1}, \ldots, x^{\gamma_m})$. Then

$\mathcal{S}_H = \{\, (m, \underline{n}, \omega) \mid M(\omega) = e_s B$ for some $s \in \{0, 1, \ldots, m\}$ where $B = J_s + A^{\underline{n}}$ for some skew-symmetric matrix $A = (a_{uv}) \in \mathfrak{Mat}_{m\times m}(F)$. If $s \geq 1$, then $a_{uv} = 0$ unless $s \in \{u,v\}$ and $\{u,v\} \not\subseteq \{s,\tilde{s}\}\,\}$.

Then our main result, which in view of Kac's isomorphism theorem implies Theorem 1, is

<u>Theorem 2.</u> Every element in $\mathcal{T}_H$ is equivalent under $\sim$ to an element of $\mathcal{S}_H$.

Our proof of this result depends on a sequence of lemmas. In Lemma 1 we use information given in ([K], Sec. 2) to show in case (b) of Theorem 1 that we may assume $M(\omega) \in e_s\mathfrak{Mat}_{m\times m}(\mathfrak{Q}(m:\underline{n}))$ for some $s \geq 1$. In the corollary to Lemma 3 and in Lemma 4 we show that this assumption can be strengthened to $M(\omega) \in e_s(J_s + \mathfrak{Mat}_{m\times m}(\mathfrak{Q}(m:\underline{n})_1)$. The corollary to Lemma 3 establishes this result for case (a) while Lemma 4 deals with case (b). In Lemmas 5, 6, and 7 we prove that in case (a) we may assume that the only nonzero, nonscalar term in each entry ω_{ij} is a scalar multiple (possibly zero) of $x^{\gamma_{ij}}$, and that in case (b) $M(\omega)$ is e_s times the sum of J_s and a matrix whose $(i\,j)$ entry is a scalar multiple (possibly zero) of $x^{\gamma_{ij}}$. As with the preceding pair of lemmas, Lemma 5 treats case (a) while Lemmas 6 and 7 deal with case (b). Lemma 8 shows that in case (b) we can assume that $\omega_{s\tilde{s}} = -\omega_{\tilde{s}s} = \pi(s)e_s$; i.e. no multiple of $x^{\gamma_{s\tilde{s}}}$ occurs as a summand of $\omega_{s\tilde{s}}$

or $\omega_{\tilde{s}s}$.

The divided power algebra $\mathcal{O}(m)$ decomposes into the direct sum $\mathcal{O}(m)_1 \cdot \mathcal{O}(m)_1 \oplus M$ where M is the span of $\{x^{p^u\varepsilon_v} \mid u \geq 0, 1 \leq v \leq m\} \cup \{x^0 = 1\}$. Throughout this paper we let λ be the projection of $\mathcal{O}(m)$ onto M, and we set $N(m:\underline{n}) = \ker \lambda|_{\mathcal{O}(m:\underline{n})} = \text{span}\{x^\alpha \mid \alpha \in A(m:\underline{n}),\ \alpha \neq 0 \text{ or } p^u\varepsilon_v \text{ for any } u, v\}$. We use λ_{uv} to denote the projection of $\mathcal{O}(m:\underline{n})$ onto $Fx^{p^u\varepsilon_v}$. Note that $\lambda_{uv}\lambda = \lambda_{uv}$.

<u>Lemma 1.</u> Let $f = \Sigma_{\alpha \in A(m)} f_\alpha x^\alpha \in \mathcal{O}(m)$. Suppose $f_0 = 1$ and $f^{-1}D_i f \in \mathcal{O}(m:\underline{n})$ for $i = 1,\ldots,m$. Then either $f \in \mathcal{O}(m:\underline{n})$ or there is an automorphism $\varphi \in \text{Aut}_{div}\mathcal{O}(m)$ and some r with $1 \leq r \leq m$ such that

$$\varphi f \in e_r \mathcal{O}(m:\underline{n}).$$

<u>Proof.</u> Suppose that $f \notin \mathcal{O}(m:\underline{n})$. Then there exists $\alpha \in A(m)$, $\alpha \notin A(m:\underline{n})$ such that $f_\alpha \neq 0$ and such that $f_\beta = 0$ for all $\beta \in A(m)$ with $\beta \notin A(m:\underline{n})$ and $|\beta| < |\alpha| = \alpha_1 + \ldots + \alpha_n$. It follows that $f \in \mathcal{O}(m:\underline{n}) + \mathcal{O}(m)_{|\alpha|}$. Since by assumption $f^{-1}D_i f \in \mathcal{O}(m:\underline{n})$ for $i = 1,\ldots,m$, we have

$$D_i f \in f\mathcal{O}(m:\underline{n}) \subseteq (\mathcal{O}(m:\underline{n}) + \mathcal{O}(m)_{|\alpha|})\mathcal{O}(m:\underline{n}) \subseteq \mathcal{O}(m:\underline{n}) + \mathcal{O}(m)_{|\alpha|}.$$

Thus, $x^{\alpha - \varepsilon_i} \in \mathcal{O}(m:\underline{n})$ for all i, so that $\alpha - \varepsilon_i \in A(m:\underline{n})$ for all i with $\alpha(i) > 0$. This implies that $\alpha = p^{n_j}\varepsilon_j$ for some j. Consequently, if we set $f_i = f_{p^{n_i}\varepsilon_i}$ and $R_f = \{ i \mid f_i \neq 0\}$, then we can conclude that $R_f \neq \emptyset$.

Therefore, when $f \notin \mathcal{O}(m:\underline{n})$, we can find $r \in R_f$ such that $n_i \geq n_r$ for all $i \in R_f$. Let $\eta \in \text{Aut}_{div}\mathcal{O}(m)$ be defined by $\eta x_r = f_r^{-1/p^{n_r}} x_r$, $\eta x_j = x_j$ for $j \neq r$. Clearly, $\eta(\mathcal{O}(m:\underline{n})) \subseteq \mathcal{O}(m:\underline{n})$. Write $\eta f = \Sigma_{\beta \in A(m)} f^\eta_\beta x^\beta$ where f^η_β is the coefficient of x^β in ηf. Then $f^\eta_r = f^\eta_{p^{n_r}\varepsilon_r} = 1$. Thus, by replacing f by ηf if necessary, we may assume that $f_r = 1$. Hence, we suppose that $r \in R_f$, $n_i \geq n_r$ for all $i \in R_f$ and $f_r = 1$. Let $\psi \in \text{Aut}_{div}\mathcal{O}(m)$ be defined by

$$\psi x_r = x_r - \sum_{\substack{i \in R_f \\ i \neq r}} f_i^{1/p^{n_r}} x^{(p^{n_i - n_r})\varepsilon_i},$$

$\psi x_j = x_j$ for $j \neq r$. We now compute $f_{p^{n_j}\varepsilon_j}^{\psi}$ for all j. We claim that for any $\psi \in \mathrm{Aut}_{div}\mathcal{Q}(m)$, $\lambda\psi = \lambda\psi\lambda$. Indeed, $\psi(\mathcal{Q}(m)_1 \cdot \mathcal{Q}(m)_1) \subseteq \mathcal{Q}(m)_1 \cdot \mathcal{Q}(m)_1$. Thus, if $a = a_1 + a_2$ with $a_1 \in \mathcal{Q}(m)_1 \cdot \mathcal{Q}(m)_1$ and $a_2 \in M$, then $\lambda\psi a = \lambda\psi a_1 + \lambda\psi a_2 = \lambda\psi a_2 = \lambda\psi\lambda a$ to give the assertion. Now

$$f_{p^{n_j}\varepsilon_j}^{\psi}\, x^{p^{n_j}\varepsilon_j} = \lambda_{n_j,j}\, \psi f = \lambda_{n_j,j}\, \lambda\psi f = \lambda_{n_j,j}\, \lambda\psi\lambda f$$

$$= \lambda_{n_j,j}\, \lambda\psi(1 + \Sigma_{u,v}\, f_{p^u\varepsilon_v}\, x^{p^u\varepsilon_v}) = \lambda_{n_j,j}\, \lambda\,(1 + \Sigma_{u,v}\, f_{p^u\varepsilon_v}\, (\psi x_v)^{(p^u)})$$

$$= \lambda_{n_j,j}\, \lambda \Big(1 + \Sigma_{u,\, v \neq r}\, f_{p^u\varepsilon_v}\, x^{p^u\varepsilon_v} +$$

$$\Sigma_u\, f_{p^u\varepsilon_r}\, (x_r - \sum_{\substack{i \in R_f \\ i \neq r}} f_i^{1/p^{n_r}} x^{(p^{n_i - n_r})\varepsilon_i})^{(p^u)} \Big).$$

Note that if $a,b \in \mathcal{Q}(m)_1$, then

$$\lambda\,(a+b)^{(k)} = \lambda\,(\sum_{i=0}^{k} a^{(i)} b^{(k-i)}) = \lambda\,(a^{(k)} + b^{(k)}).$$

Thus,

$$f_{p^{n_j}\varepsilon_j}^{\psi}\, x^{p^{n_j}\varepsilon_j} = \lambda_{n_j,j}\, \lambda \Big(1 + \Sigma_{u,\, v \neq r}\, f_{p^u\varepsilon_v}\, x^{p^u\varepsilon_v}$$

$$+ \Sigma_u\, f_{p^u\varepsilon_r}\, (x^{p^u\varepsilon_r} - \sum_{\substack{i \in R_f \\ i \neq r}} f_i^{p^{u-n_r}} x^{(p^{u+n_i-n_r})\varepsilon_i}) \Big).$$

For $j \neq r$ this equation reads

$$f_{p^{n_j}\varepsilon_j}^{\psi}\, x^{p^{n_j}\varepsilon_j} = f_{p^{n_j}\varepsilon_j}\, x^{p^{n_j}\varepsilon_j} - f_{p^{n_r}\varepsilon_r}\, f_j\, x^{p^{n_j}\varepsilon_j} = 0$$

because $f_{p^{n_r}\varepsilon_r} = f_r = 1$ and $f_{p^{n_j}\varepsilon_j} = f_j$. Thus we may suppose that $R_f = \{r\}$

and $f_r = 1$.

With that assumption let $g = e_r^{-1}f = \exp(-x^{p^{n_r}\varepsilon_r})f$. Clearly $R_g = \emptyset$. Observe that since $D(a^{(k)}) = D(a)\,a^{(k-1)}$, it follows that $D(\exp(a)) = \exp(a)D(a)$ for all $D \in W(m)$, $a \in \mathcal{O}(m)$. Thus, we have

$$g^{-1}D_i g = e_r f^{-1} D_i e_r^{-1} f = e_r f^{-1} D_i (e_r^{-1}) f + e_r f^{-1} e_r^{-1} D_i f$$

$$= -\delta_{ir}\, x^{(p^{n_r}-1)\varepsilon_r} + f^{-1}D_i f \in \mathcal{O}(m:\underline{n})$$

for any i. Since $R_g = \emptyset$, the first part of the proof implies that $g \in \mathcal{O}(m:\underline{n})$. Hence for $\varphi = \psi\eta$, $\varphi f \in e_r\mathcal{O}(m:\underline{n})$ as required. ■

Lemma 2. Assume $\omega = \Sigma_{i,j}\, \omega_{ij}\, dx_i \wedge dx_j$ and $\varphi \in \operatorname{Aut}\mathcal{O}(m)$. Then $M(\varphi\omega) = \mathcal{J}(\varphi)(\varphi M(\omega))\,\mathcal{J}(\varphi)^t$ where $\mathcal{J}(\varphi)$ is the Jacobian matrix $(D_i\varphi x_j)$, and φ acts on $M(\omega)$ componentwise.

Proof. The result follows readily from

$$\varphi\omega = \Sigma_{i,j}\ \varphi(\omega_{ij})\, d\varphi x_i \wedge d\varphi x_j = \Sigma_{i,j,k,\ell}\, \varphi(\omega_{ij}) D_k(\varphi x_i) D_\ell(\varphi x_j)\, dx_k \wedge dx_\ell$$

$$= \Sigma_{k,\ell}\left(\Sigma_{i,j}\ D_k(\varphi x_i)\varphi(\omega_{ij}) D_\ell(\varphi x_j)\right) dx_k \wedge dx_\ell \,. \quad ■$$

Lemma 3. Let $A_0 \in \mathfrak{Mat}_{m\times m}(F)$ be skew-symmetric and nonsingular. Then there exist a lower triangular matrix $L \in \mathfrak{Mat}_{m\times m}(F)$ and a permutation matrix $P \in \mathfrak{Mat}_{m\times m}(F)$ such that $(PL)\,A_0\,(PL)^t = J_0$.

Proof. Theorem 1 of ([H,P], Sec.3, Chap.9) gives that any such matrix A_0 is congruent by a lower triangular matrix to a matrix of the form $\Sigma_i\, \tau_i E_{i,\zeta(i)}$ where $\tau_i = -\tau_{\zeta(i)} = \pm 1$ and $\zeta \in \mathcal{S}ym(1,\dots,m)$ with $\zeta^2 = 1$. Choose $\sigma \in \mathcal{S}ym(1,\dots,m)$ such that $\sigma(\{i \mid \tau_i = 1\}) \subseteq \{1,\dots,m/2\}$ and $\sigma\zeta(i) = \widetilde{\sigma(i)}$ for all i. If $P(\sigma) = \Sigma_i\, E_{\sigma(i),i}$, then $P(\sigma)\left(\Sigma_i \tau_i E_{i,\zeta(i)}\right)P(\sigma)^t = \Sigma_i\, \tau_i E_{\sigma(i),\sigma\zeta(i)} = J_0$ as desired. ■

Corollary. Let $A \in \mathfrak{Mat}_{m\times m}(\mathcal{O}(m))$ be skew-symmetric and nonsingular. Then there exist $\varphi \in \operatorname{Aut}_{div}\mathcal{O}(m)$ such that $\varphi x_i = \Sigma_{j\geq i}\, b_{ji} x_j$ for all i where $b_{ji} \in F$ for all i,j, and a permutation $\sigma \in \mathcal{S}ym(1,\dots,m)$ with associated automorphism $\sigma \in \operatorname{Aut}_{div}\mathcal{O}(m)$ given by $\sigma x_i = x_{\sigma(i)}$ for all i, such that

$$\mathcal{J}(\sigma\varphi)(\sigma\varphi A)\mathcal{J}(\sigma\varphi)^t \equiv \mathcal{J}(\sigma\varphi) A_0 \mathcal{J}(\sigma\varphi)^t \equiv J_0 \quad \text{mod } \mathfrak{Mat}_{m\times m}(\mathcal{Q}(m)_1).$$

where A_0 is the scalar part of A.

Proof. Write $A = A_0 + A_1$ where $A_0 \in \mathfrak{Mat}_{m\times m}(F)$ and $A_1 \in \mathfrak{Mat}_{m\times m}(\mathcal{Q}(m)_1)$. Choose $L = (b_{uv})$ and $P = P(\sigma)$ for $\sigma \in \mathcal{S}ym(1,\ldots,m)$ as in Lemma 3. Define φ and σ as in the statement of this corollary. Since automorphisms preserve F and $\mathcal{Q}(m)_1$, and since $\mathcal{J}(\sigma\varphi) = \mathcal{J}(\sigma)\mathcal{J}(\varphi)$, where $\mathcal{J}(\sigma) = P$, and $\mathcal{J}(\varphi) = L$, the assertion is immediate. ■

Lemma 4. If $(m, \underline{n}, \omega)$ satisfies the compatibility condition (equation (2)), then there is a triple $(m, \underline{n}', \omega') \sim (m, \underline{n}, \omega)$ satisfying the compatibility condition with $M(\omega') \in e'_s(J_s + \mathfrak{Mat}_{m\times m}(\mathcal{Q}(m:\underline{n}')_1)$, for some s, $0 \le s \le m$, where $e'_s = \exp\left(x^{p^{n'_s}\varepsilon_s}\right)$.

Proof. By Lemma 1 we may suppose that $M(\omega) = e_r B$ for some r, $0 \le r \le m$, where $B \in \mathfrak{Mat}_{m\times m}(\mathcal{Q}(m:\underline{n}))$. By applying a permutation automorphism we may assume without loss of generality that $n_1 \le \ldots \le n_m$, and if $r \neq 0$, that either $n_r < n_{r+1}$ or $r = m$. Applying the corollary to Lemma 3 with $A = M(\omega)$ produces $\varphi \in \mathrm{Aut}_{\mathrm{div}}\mathcal{Q}(m)$ with $\varphi(\mathcal{Q}(m:\underline{n})) \subseteq \mathcal{Q}(m:\underline{n})$, and $\sigma \in \mathcal{S}ym(1,\ldots,m)$, with $\sigma \in \mathrm{Aut}_{\mathrm{div}}\mathcal{Q}(m)$ given by $\sigma x_i = x_{\sigma(i)}$ for all i such that $\mathcal{J}(\sigma\varphi)(\sigma\varphi M(\omega))\mathcal{J}(\sigma\varphi)^t \equiv J_0$ mod $\mathfrak{Mat}_{m\times m}(\mathcal{Q}(m)_1)$. By Lemma 2,

$$\begin{aligned} M(\sigma\varphi\omega) &= \mathcal{J}(\sigma\varphi)\big(\sigma\varphi M(\omega)\big)\,\mathcal{J}(\sigma\varphi)^t = \mathcal{J}(\sigma\varphi)\big(\sigma\varphi e_r B\big)\mathcal{J}(\sigma\varphi)^t \\ &= (\sigma\varphi e_r)\left(\mathcal{J}(\sigma\varphi)(\sigma\varphi B)\,\mathcal{J}(\sigma\varphi)^t\right) \end{aligned}$$

If $r \ge 1$, then $\varphi e_r = \exp\left((\varphi x_r)^{(p^{n_r})}\right) = \exp\left((\Sigma_{j\ge r} b_{jr} x_j)^{(p^{n_r})}\right)$. Since $r = m$ or $n_j > n_r$ for all $j > r$, we have that $(\Sigma_{j\ge r} b_{jr} x_j)^{(p^{n_r})} \equiv b_{rr}^{p^{n_r}} x^{p^{n_r}\varepsilon_r}$ mod $\mathcal{Q}(m:\underline{n})$. Hence $\exp\left((\Sigma_{j\ge r} b_{jr} x_j)^{(p^{n_r})}\right) \in \exp\left(b_{rr}^{p^{n_r}} x^{p^{n_r}\varepsilon_r}\right)\mathcal{Q}(m:\underline{n})$. Let $\underline{n}' = (n'_1,\ldots,n'_m)$ where $n'_i = n_{\sigma^{-1}(i)}$. Then $\sigma\colon \mathcal{Q}(m:\underline{n}) \longrightarrow \mathcal{Q}(m:\underline{n}')$, and $\sigma\varphi e_r \in \sigma\left(\exp\left(b_{rr}^{p^{n_r}} x^{p^{n_r}\varepsilon_r}\right)\mathcal{Q}(m:\underline{n})\right) \subseteq \exp\left(b_{rr}^{p^{n_r}} x^{p^{n_r}\varepsilon_{\sigma(r)}}\right)\mathcal{Q}(m:\underline{n}')$. Writing $s = \sigma(r)$

we have $M(\sigma\varphi\omega) \in \exp\left(b_{rr}^{p^{n'_s}} x^{p^{n'_s}\varepsilon_s}\right) \mathfrak{M}at_{m\times m}(\mathcal{O}(m:\underline{n}'))$. By applying an automorphism η which takes x_s to $b_{rr}^{-1}x_s$ and $x_{\tilde{s}}$ to $b_{rr}x_{\tilde{s}}$ and fixes x_j for $j \neq s,\tilde{s}$, we have that $M(\omega') \equiv \exp\left(x^{p^{n'_s}\varepsilon_s}\right) J_0 \mod \mathfrak{M}at_{m\times m}(\mathcal{O}(m:\underline{n}')_1)$ where $\omega' = \eta\sigma\varphi\omega$. Since $VU_s^t - U_sV \in \mathfrak{M}at_{m\times m}(\mathcal{O}(m:\underline{n}')_1)$, we may write $M(\omega') \equiv \exp\left(x^{p^{n'_s}\varepsilon_s}\right) J_s \mod \mathfrak{M}at_{m\times m}(\mathcal{O}(m:\underline{n}')_1)$.

It remains to show that $(m,\underline{n}',\omega') \sim (m,\underline{n},\omega)$. Since $\varphi \in \mathrm{Aut}_{div}\mathcal{O}(m)$ and $\varphi(\mathcal{O}(m:\underline{n})) \subseteq \mathcal{O}(m:\underline{n})$, we have $(m,\underline{n},\omega) \sim (m,\underline{n},\varphi\omega)$. But $W(m:\underline{n}) \cap \{D \in W(m) \mid D(\varphi\omega) = 0\}$ maps isomorphically under σ to $\sigma W(m:\underline{n})\sigma^{-1} \cap \{D \in W(m) \mid D(\sigma\varphi\omega) = 0\} = W(m:\underline{n}') \cap \{D \in W(m) \mid D(\sigma\varphi\omega) = 0\}$, and $W(m:\underline{n}') \cap \{D \in W(m) \mid D(\sigma\varphi\omega) = 0\}$ maps isomorphically under η to $W(m:\underline{n}') \cap \{D \in W(m) \mid D(\eta\sigma\varphi\omega) = 0\}$ so that $(m,\underline{n}',\omega') \sim (m,\underline{n},\omega)$. The compatibility condition holds for $M(\omega')$ by its construction. ∎

<u>Lemma 5.</u> Let $C = (c_{ij}) \in \mathfrak{M}at_{m\times m}(\mathcal{O}(m:\underline{n}))$ be skew-symmetric. Assume that $D_i c_{jk} + D_j c_{ki} + D_k c_{ij} = 0$ for all i,j,k and that $D_i^{p^{n_i}-1} D_j^{p^{n_j}-1} c_{ij} = 0$ for all i,j. Then there exist $a_1,\ldots,a_m \in N(m:\underline{n})$ such that $c_{ij} = D_i a_j - D_j a_i$ for all i,j.

<u>Proof.</u> By Lemma 9 of [W,2] we may find $a_1',\ldots,a_m' \in \mathcal{O}(m:\underline{n})$ such that $c_{ij} = D_i a_j' - D_j a_i'$ for all i,j. We will show that there exists $b \in \mathcal{O}(m:\underline{n})$ such that $a_i' - D_i b \in N(m:\underline{n})$ for all i. Then setting $a_i = a_i' - D_i b$ gives the desired result.

Let $a_i' = \sum_{\alpha\in A(m:\underline{n})} a_{i,\alpha} x^{\alpha}$ and set

$$b = \sum_{i,v=1}^{m} \sum_{u=0}^{n_v-1} a_{i,\,p^u\varepsilon_v}\, x^{p^u\varepsilon_v+\varepsilon_i}$$

Then we must show that $\lambda(a_j' - D_j b) = 0$ for all j. But

$$\lambda\left(D_j \sum_{i,v=1}^{m} \sum_{u=0}^{n_v-1} a_{i,\,p^u\varepsilon_v}\, x^{p^u\varepsilon_v+\varepsilon_i}\right) = \sum_{v=1}^{m} \sum_{u=0}^{n_v-1} a_{j,\,p^u\varepsilon_v}\, x^{p^u\varepsilon_v} = \lambda a_j'. \ ∎$$

<u>Lemma 6.</u> Let $a_1,\ldots,a_m \in N(m:\underline{n})_t$ where $t > 0$. Then there exists $\varphi \in \mathrm{Aut}_{div}\mathcal{O}(m:\underline{n})$ such that $\varphi x_i = x_i + a_i$ for all i. For any j with $1 \le j \le m$, $\varphi(e_j)$

$\equiv e_j \mod \mathcal{O}(m:\underline{n})_{p^{n_j}+t-1}$.

Proof. That such a φ exists follows from Corollary 1 to Lemma 9 of [W,1]. Then we have

$$\varphi(e_j) = \exp((\varphi x_j)^{(p^{n_j})}) = \exp((x_j + a_j)^{(p^{n_j})}) = \exp(\sum_{i=0}^{p^{n_j}} x_j^{(i)} a_j^{(p^{n_j}-i)}).$$

But $a_j^{(k)} \in \mathcal{O}(m:\underline{n})$ for all k by ([W,1] Lemma 10), and $x_j^{(i)} \in \mathcal{O}(m:\underline{n})$ for $i < p^{n_j}$ so that

$$\varphi(e_j) = e_j \exp(\sum_{i=0}^{p^{n_j}-1} x_j^{(i)} a_j^{(p^{n_j}-i)}) \equiv e_j \mod \mathcal{O}(m:\underline{n})_{p^{n_j}+t-1}$$

as required. ■

Definition. Let $B = (b_{ij}) \in \mathfrak{Mat}_{m\times m}(\mathcal{O}(m:\underline{n}))$ and write

$$b_{ij} = \Sigma_{\gamma \in A(m:\underline{n})}\ b_{ij,\gamma}\ x^{\gamma}$$

where $b_{ij,\gamma} \in F$. Then $B^* = (b_{ij}^*) \in \mathfrak{Mat}_{m\times m}(\mathcal{O}(m:\underline{n}))$ is the matrix having $b_{ij}^* = b_{ij,\gamma_{ij}} x^{\gamma_{ij}}$ where γ_{ij} is as above.

Theorem 1 asserts that we may assume $M(\omega) = e_s J_s + M(\omega)^*$ for some $s \geq 0$.

Lemma 7. Let $M(\omega) = e_s B$ for some s, $0 \leq s \leq m$, and $B \in \mathfrak{Mat}_{m\times m}(\mathcal{O}(m:\underline{n}))$. Assume $M(\omega)$ satisfies the Hamiltonian condition (equation (1)). Suppose that

$$B \equiv J_s + B^* \mod \mathfrak{Mat}_{m\times m}(\mathcal{O}(m:\underline{n})_i),\ \ i > 0.$$

Then there exist elements $a_1, \ldots, a_m \in N(m:\underline{n})_{[i]}$ such that if $\varphi \in \mathrm{Aut}_{div}\mathcal{O}(m:\underline{n})$ is as in Lemma 6, then $M(\varphi\omega) = e_s B'$ where $B' \equiv J_s + (B')^* \mod \mathfrak{Mat}_{m\times m}(\mathcal{O}(m:\underline{n})_{i+1})$. Also $B^* \equiv (B')^* \mod \mathfrak{Mat}_{m\times m}(\mathcal{O}(m:\underline{n})_{i+1})$.

Proof. Let $B - B^* \equiv J_s + (c_{uv}) \mod \mathfrak{Mat}_{m\times m}(\mathcal{O}(m:\underline{n})_{i+1})$ where $c_{u,v} \in \mathcal{O}(m:\underline{n})_{[i]}$ for all u,v. We have $c_{uv} = -c_{vu}$ for all u,v. Since the (ij) entry of B^* is a scalar multiple of $x^{\gamma_{ij}}$, B^* always satisfies the Hamiltonian condition. Therefore, $D_u c_{vw} + D_v c_{wu} + D_w c_{uv} = 0$ for all u,v,w if $s = 0$, and for all u,v,w with $s \notin \{u,v,w\}$ if $s \geq 1$. We suppose that $s \geq 1$ and consider $D_s c_{uv} + D_u c_{vs} + D_v c_{su}$. It is sufficient to let (d_{uv}) denote the degree i component of $B - J_s$ and to show that $D_s d_{uv} + D_u d_{vs} + D_v d_{su} = 0$. The Hamiltonian condition for $e_s B$ gives $D_s(e_s b_{uv}) + D_u(e_s b_{vs}) + D_v(e_s b_{su}) = 0$ so that $e_s^{-1}(D_s(e_s b_{uv}) + D_u(e_s b_{vs}) + D_v(e_s b_{su})) = 0$. Hence,

$$x^{(p^{n_s}-1)\varepsilon_s} b_{uv} + D_s(b_{uv}) + D_u(b_{vs}) + D_v(b_{su}) = 0.$$

The Hamiltonian condition for $e_s J_s$ where $J_s = (j_{uv})$ implies

$$x^{(p^{n_s}-1)\varepsilon_s} j_{uv} + D_s(j_{uv}) + D_u(j_{vs}) + D_v(j_{su}) = 0.$$

Hence, letting $k_{uv} = b_{uv} - j_{uv}$ (so that d_{uv} is the degree i component of k_{uv}) we get

$$x^{(p^{n_s}-1)\varepsilon_s} k_{uv} + D_s(k_{uv}) + D_u(k_{vs}) + D_v(k_{su}) = 0.$$

Thus, $D_s d_{uv} + D_u d_{vs} + D_v d_{su}$ is $-x^{(p^{n_s}-1)\varepsilon_s}$ times the degree $i - p^{n_s}$ component of k_{uv}.

By assumption $B - J_s \equiv B^*$ mod $\mathfrak{Mat}_{m\times m}(\mathcal{Q}(m:\underline{n})_i)$, so any nonzero component of k_{uv} of degree less than i must be of degree $p^{n_u} + p^{n_v} - 2$ and must be a multiple of $x^{\gamma_{uv}}$. Since no term in $D_s d_{uv} + D_u d_{vs} + D_v d_{su}$ can contain $x^{\gamma_{uv}} x^{(p^{n_s}-1)\varepsilon_s}$, we must have that this sum is zero and that the Hamiltonian condition holds for the d_{uv} and hence for the c_{uv}.

Now by Lemma 5 there exist $\grave{a}_1, \ldots, \grave{a}_m \in N(m:\underline{n})$ such that $c_{uv} = D_v \grave{a}_u - D_u \grave{a}_v$ for all u,v. Since $c_{uv} \in \mathcal{Q}(m:\underline{n})_{[i]}$ for all u,v, we may suppose that $\grave{a}_u \in N(m:\underline{n})_{[i+1]}$ for all u. Letting $a_1, \ldots, a_m$ be given so that $-\pi(u)a_{\tilde{u}} = \grave{a}_u$, we have $-c_{uv} = D_v \pi(u) a_{\tilde{u}} - D_u \pi(v) a_{\tilde{v}}$. Assume that φ is the automorphism of $\mathcal{Q}(m:n)$ defined by $\varphi x_j = x_j + a_j$ for all j (as discussed in Lemma 6). Then

$$M(\varphi\omega) = \mathcal{J}(\varphi)\varphi(e_s B)\mathcal{J}(\varphi)^t = (\varphi e_s)\ \mathcal{J}(\varphi)\ (\varphi B)\ \mathcal{J}(\varphi)^t.$$

Now $\varphi e_s \equiv e_s \bmod \mathcal{Q}(m:\underline{n})_{i+1}$ by Lemma 6, so it suffices to show that $\mathcal{J}(\varphi)(\varphi B)\mathcal{J}(\varphi)^t \equiv J_s + B^* \bmod \mathfrak{Mat}_{m\times m}(\mathcal{Q}(m:\underline{n})_{i+1})$. But $\varphi B \equiv B \bmod \mathfrak{Mat}_{m\times m}(\mathcal{Q}(m:\underline{n})_{i+1})$ and $\mathcal{J}(\varphi) = (D_u\varphi x_v) = (D_u(x_v + a_v)) = I + (D_u a_v)$. Therefore,

$$\mathcal{J}(\varphi)\ (\varphi B)\ \mathcal{J}(\varphi)^t \equiv (I + (D_u a_v))\ B\ (I + (D_u a_v)^t) \equiv$$

$$\equiv B + (D_u a_v) J_0 + J_0 (D_u a_v)^t \mod \mathfrak{Mat}_{m\times m}(\mathcal{Q}(m:\underline{n})_{i+1}).$$

Thus modulo $\mathcal{Q}(m:\underline{n})_{i+1}$, the (u,v) entry in $\mathcal{J}(\varphi)\ (\varphi B)\ \mathcal{J}(\varphi)^t$ is $b_{uv} + D_v\pi(u)a_{\tilde{u}} - D_u\pi(v)a_{\tilde{v}}$ which is $b_{uv} - c_{uv}$. So $B' \equiv J_s + (B')^* \bmod \mathfrak{Mat}_{m\times m}(\mathcal{Q}(m:\underline{n})_{i+1})$, as required. ∎

<u>Lemma 8.</u> Suppose that $M(\omega) = e_s B$ where $B = J_s + B^*$, $B^* = (b_{uv})$ with $b_{uv} \in F x^{\gamma_{uv}}$. Then replacing ω by $\varphi\omega$ for some $\varphi \in \mathrm{Aut}_{div}\mathcal{Q}(m:\underline{n})$ if necessary, we may assume that $b_{s\tilde{s}} = b_{\tilde{s}s} = 0$.

<u>Proof.</u> Define $\psi \in \mathrm{Aut}_{div}\mathcal{Q}(m:\underline{n})$ by $\psi x_s = x_s - \pi(s)b_{s\tilde{s}}x^{\gamma_{\tilde{s}}}$, $\psi x_u = x_u$ for $u \neq s$. Then

$$\psi e_s = \exp((x_s - \pi(s)b_{s\tilde{s}}x^{\gamma_{\tilde{s}}})^{(p^{n_s})}) = \exp\Big(\sum_{j=0}^{p^{n_s}} (x_s)^{(j)}(-\pi(s)b_{s\tilde{s}}x^{\gamma_{\tilde{s}}})^{(p^{n_s}-j)}\Big)$$

$$= \exp(x^{p^{n_s}\varepsilon_s} - x^{\gamma_s}\pi(s)\ b_{s\tilde{s}}x^{\gamma_{\tilde{s}}}) = e_s \exp(-\pi(s)b_{s\tilde{s}}x^{\gamma_{s\tilde{s}}}).$$

Thus,

$$M(\psi\omega) = \psi(e_s)\mathcal{J}(\psi)\ (\psi B)\ \mathcal{J}(\psi)^t$$

$$= e_s(1-\pi(s)b_{s\tilde{s}}x^{\gamma_{s\tilde{s}}})(I-\pi(s)b_{s\tilde{s}}x^{\gamma_{\tilde{s}}-\varepsilon_{\tilde{s}}}E_{s\tilde{s}})\psi B(I-\pi(s)b_{s\tilde{s}}x^{\gamma_{\tilde{s}}-\varepsilon_{\tilde{s}}}E_{s\tilde{s}})^t,$$

where $E_{s\tilde{s}}$ represents the standard matrix unit. Now let $i = |\gamma_s + \gamma_{\tilde{s}}|$. Clearly $\psi B^* \equiv B^* \bmod \mathfrak{Mat}_{m\times m}(\mathcal{Q}(m:\underline{n})_{i+1})$. Also $\psi J_s \equiv J_s \bmod \mathfrak{Mat}_{m\times m}(\mathcal{Q}(m:\underline{n})_{i+1})$. Therefore modulo $\mathfrak{Mat}_{m\times m}(\mathcal{Q}(m:\underline{n})_{i+1})$,

$$M(\psi\omega) \equiv e_s(1-\pi(s)b_{s\tilde{s}}x^{\gamma_{s\tilde{s}}})\left(I-\pi(s)b_{s\tilde{s}}x^{\gamma_{\tilde{s}}-\varepsilon_{\tilde{s}}}E_{s\tilde{s}}\right) B \left(I-\pi(s)b_{s\tilde{s}}x^{\gamma_{\tilde{s}}-\varepsilon_{\tilde{s}}}E_{s\tilde{s}}\right)^t.$$

But $(I - aE_{s\tilde{s}})\, J_0\, (I - aE_{s\tilde{s}})^t = J_0$ for any $a \in \mathcal{O}(m)$, so that

$$(I-\pi(s)b_{s\tilde{s}}x^{\gamma_{\tilde{s}}-\varepsilon_{\tilde{s}}}E_{s\tilde{s}})\, J_s\, (I-\pi(s)b_{s\tilde{s}}x^{\gamma_{\tilde{s}}-\varepsilon_{\tilde{s}}}E_{s\tilde{s}})^t \equiv J_s$$

mod $\mathfrak{Mat}_{m\times m}(\mathcal{O}(m:\underline{n})_1)$, and

$$(I-\pi(s)b_{s\tilde{s}}x^{\gamma_{\tilde{s}}-\varepsilon_{\tilde{s}}}E_{s\tilde{s}})\, B^*\, (I-\pi(s)b_{s\tilde{s}}x^{\gamma_{\tilde{s}}-\varepsilon_{\tilde{s}}}E_{s\tilde{s}})^t \equiv B^*$$

mod $\mathfrak{Mat}_{m\times m}(\mathcal{O}(m:\underline{n})_1)$. Thus, $M(\psi\omega) \equiv e_s(1-\pi(s)b_{s\tilde{s}}x^{\gamma_{s\tilde{s}}})\, B$ mod $\mathfrak{Mat}_{m\times m}(\mathcal{O}(m:\underline{n})_1)$. Consequently, modulo $\mathfrak{Mat}_{m\times m}(\mathcal{O}(m:\underline{n})_1)$ the $(s,\tilde{s})$ entry of $M(\psi\omega)$ is $e_s(1-\pi(s)b_{s\tilde{s}}x^{\gamma_{s\tilde{s}}})(\pi(s) + b_{s\tilde{s}}x^{\gamma_{s\tilde{s}}}) = \pi(s)e_s$. Now apply the previous lemma to obtain the result. ■

Hence, we have proved the following theorem which is a reformulation of Theorem 2:

<u>Theorem 2'.</u> Assume that $M(\omega)$ satisfies the Hamiltonian and compatability conditions.

(a) If $M(\omega) \in \mathfrak{Mat}_{m\times m}(\mathcal{O}(m:\underline{n}))$, then there exists $\varphi \in \mathrm{Aut}_{\mathrm{div}}\mathcal{O}(m)$ such that $M(\varphi\omega) - J_0 = (a_{uv})$ where $a_{uv} = -a_{vu} \in Fx^{\gamma_{uv}}$.

(b) If $M(\omega) \notin \mathfrak{Mat}_{m\times m}(\mathcal{O}(m:\underline{n}))$, then there exists $s \in \{1,\ldots,m\}$ and $\varphi \in \mathrm{Aut}_{\mathrm{div}}\mathcal{O}(m)$ such that $M(\varphi\omega) - e_sJ_s = e_s(a_{uv})$, a_{uv} as above. Furthermore, in this case, $a_{uv} = 0$ unless $s \in \{u,v\}$ and $\{u,v\} \not\subseteq \{s,\tilde{s}\}$.

<u>Proof.</u> Only the last assertion needs to be verified. For that apply, as in the proof of Lemma 8, the Hamiltonian condition with one of the indices equal to s to $e_s(a_{uv})$ to obtain

$$-x^{\gamma_s}a_{uv} = D_s a_{uv} + D_u a_{vs} + D_v a_{su}.$$

The left hand side is a multiple of $x^{\gamma_s+\gamma_u+\gamma_v}$, but the right side is in $D_s\mathfrak{a}(m:\underline{n}) + D_u\mathfrak{a}(m:\underline{n}) + D_v\mathfrak{a}(m:\underline{n})$. Thus, both sides must be zero, which implies that $a_{uv} = 0$. ■

References

[H,P] W.V.D. Hodge and D. Pedoe, *Methods of Algebraic Geometry* Vol. I, Cambridge Univ. Press, New York (1947).

[K] V.G. Kac, Description of filtered Lie algebras with which graded Lie algebras of Cartan type are associated, Izv. Akad. Nauk SSSR Ser. Mat. 38 (1974), 800-834; English transl. Math. USSR-Izv. 8 (1974), 801-835. Errata 10 (1976), 1339.

[K,K] M. I. Kuznetsov and S. A. Kirillov, Hamiltonian differential forms over an algebra of truncated polynomials, Uspekhi Mat 41 (1986) no.2 (248), 197-198; English transl. Russian Math. Surveys 41 (1986), 205-206.

[T] S. A. Tyurin, The classification of deformations of a special Lie algebra Cartan type, Mat. Zametki 24 (1978), 847-857; English transl. Math. Notes 24 (1978) 948-954.

[W,1] R. L. Wilson, Classification of generalized Witt algebras over algebraically closed fields, Trans. Amer. Math. Soc. 153 (1971), 191-210.

[W,2] R.L. Wilson, Automorphisms of graded Lie algebras of Cartan type, Comm. in Algebra 7 (1975), 591-613.

[W,3] R. L. Wilson, A structural characterization of the simple Lie algebras of generalized Cartan type over fields of prime characteristic, J. Algebra 40 (1976), 418-465.

[W,4] R.L. Wilson, Simple Lie algebras of type S, J. Algebra 62 (1980), 292-298.

1980 Mathematics Subject Classification: 17B50, 17B20

G. M. Benkart and J. M. Osborn
Department of Mathematics
University of Wisconsin
Madison, Wisconsin 53706

T. B. Gregory
Department of Mathematics
Ohio State University at Mansfield
Mansfield, Ohio 44906

H. Strade
Department of Mathematics
University of Hamburg
Bundesstrasse 55
2 Hamburg 13, FR Germany

R. L. Wilson
Department of Mathematics
Rutgers University
New Brunswick, New Jersey 08903

ON LIE ALGEBRAS WITH A SUBALGEBRA OF CODIMENSION ONE

Alberto Elduque*

Let F be a field of characteristic $p > 2$. This will be assumed throughout the paper. By $W(1{:}m;\Gamma)$, where $\Gamma = (\gamma_0,\dots,\gamma_{m-1})$ is a set of scalars, we shall denote the Lie algebra over F with a basis $\{v_i : i = -1,0,\dots,p^m-2\}$ (which will be called a standard basis) and multiplication given by

$$[v_{-1},v_i] = v_{i-1} \qquad \text{if } i \neq p^j-1$$

$$[v_{-1},v_{p^j-1}] = v_{p^j-2} + \gamma_j v_{p^m-2}$$

$$[v_i,v_j] = \alpha_{ij} v_{i+j} \qquad \text{for } i,j \geq 0,$$

where $\alpha_{ij} = \binom{i+j+1}{j} - \binom{i+j+1}{i}$ and $v_h = 0$ if $h > p^m-2$.

These algebras are simple and contain a subalgebra of codimension one, the span of $\{v_i : i = 0,\dots,p^m-2\}$. If $\gamma_j = 0$ for all j, then $W(1{:}m;\Gamma) = W(1{:}m)$ is the Zassenhaus algebra of dimension p^m.

Lie algebras with a subalgebra of codimension one were studied by Amayo as a part of his study of the so called quasiideals of a Lie algebra. He proved ([1; Theorem 3.1]) that a finite dimensional Lie algebra over F has a subalgebra of codimension one containing no nonzero ideals of the algebra if and only if it is either one or two dimensional, isomorphic to sl(2,F) or isomorphic to some $W(1{:}m;\Gamma)$. Then he proved ([1; Theorem 3.9]) that all the algebras $W(1{:}m;\Gamma)$ were isomorphic.

This last result was shown to be false in general by Benkart, Isaacs and Osborn ([2]), who gave examples over nonperfect fields of nonisomorphic Lie algebras of dimension p^2 with subalgebras of codimension one. They also proved that all these algebras $W(1{:}m;\Gamma)$ are in fact isomorphic over algebraically closed fields.

Later on, Dzhumadil'daev ([5]) and Varea ([7]) showed that these $W(1{:}m;\Gamma)$ are also isomorphic over perfect fields, the key is the existence in these algebras of ad–nilpotent elements which do not belong to the subalgebra of codimension one.

In this paper we shall use the results in [2] and [3] on the Zassenhaus algebras to give a new proof of the classification of the Lie algebras with a subalgebra of codimension one containing no nonzero ideals, given in [1; Theorem 3.1]. We shall give conditions for two algebras $W(1{:}m;\Gamma)$ to be isomorphic, showing that there are infinitely many classes of isomorphism when $m \geq 2$ and F is a nonperfect field. Finally, we shall study the automorphism group of all these algebras and give a concrete description of it by means of some exponentials. In particular we shall describe the automorphism group of the Zassenhaus algebras, which has been studied by Ree ([6]) and Wilson ([8]).

---

* Partially supported by the CAICYT (PR. 84-0778).

All the algebras considered will be finite dimensional over F.

1. First results

We shall use the following known results on the Zassenhaus algebra of dimension p^m. Their proofs may be found in [2] and [3]:

Theorem 1.1

a) If $p^m > 3$ ($p^m = 3$ implies that $W(1{:}m) \cong sl(2,F)$) then $W(1{:}m)$ has a unique subalgebra of codimension one.

b) The characteristic polynomial of $\mathrm{ad}\,y$, $y \in W(1{:}m) - S$, S the subalgebra of codimension one, is equal to the minimal polynomial and has the form

$$X^{p^m} + \beta_{m-1}X^{p^{m-1}} + \ldots + \beta_0 X$$

where $\beta_i \in F^{p^i}$ for all i.

Moreover, if $y = v_{-1} + \sum_{i=1}^{m} \alpha_i v_{p^i-2}$, then the minimal polynomial of $\mathrm{ad}\,y$ is

$$h_m(X) = X^{p^m} - (\alpha_1 X)^{p^{m-1}} - \ldots - (\alpha_{m-1}X)^p - \alpha_m X.$$

c) The centralizer of every y in $W(1{:}m) - S$ is Fy.

d) If F is algebraically closed and L is a Lie algebra with a subalgebra of codimension one which contains no nonzero ideal of L, then either $\dim L \leq 2$, $L \cong sl(2,F)$ or $L \cong W(1{:}m)$ for some m.

Let L be a Lie algebra over F and S a subalgebra of codimension one. We construct the usual filtration in L as follows: $L_{-1} = L$, $L_0 = S$ and $L_{i+1} = \{x \in L_i : [x,L] \subseteq L_i\}$ for $i \geq 0$.

Let r be the smallest nonnegative integer with $L_r = L_{r+1}$. Then L_r is the largest ideal of L contained in S, the L_i's ($i \geq 0$) are ideals of S and, if $x \in L-S$, then $L_{i+1} = \{z \in S : [x,z] \in L_i\}$ and $\dim(L_i / L_{i+1}) = 1$ for $i < r$.

Let Ω be the algebraic closure of F. The associated filtration in L_Ω with respect to S_Ω is $\{(L_i)_\Omega\}_{i=-1}^{r}$. Hence $(L_r)_\Omega = (L_\Omega)_r$ is the largest ideal of L_Ω contained in S_Ω.

Lemma 1.2 Let L be a Lie algebra over F with a subalgebra S of codimension one containing no nonzero ideal of L. Then either $\dim L \leq 2$, L is isomorphic to $sl(2,F)$ or L is a form of a Zassenhaus algebra.

Proof: L_Ω also has a subalgebra of codimension one containing no nonzero ideal of L_Ω. Now we apply Theorem 1.1 d).

Let us suppose now that L is a form of a Zassenhaus algebra of dimension p^m and S is a subalgebra of codimension one (unique if $p^m > 3$). Let x be an element in L–S. From Theorem 1.1 b) the minimal polynomial of adx is

$$X^{p^m} - \gamma_{m-1}X^{p^{m-1}} - \ldots - \gamma_0 X$$

where $\gamma_0,\ldots,\gamma_{m-1} \in F$.

Theorem 1.3 With L, S and $(\gamma_0,\ldots,\gamma_{m-1}) = \Gamma$ as above, L is isomorphic to $W(1{:}m;\Gamma)$.

Proof: Let $\{L_i\}_{i=-1}^{p^m-2}$ be the associated filtration of L and take $z_{p^m-2} \neq 0$ in L_{p^m-2}. We define recursively $z_{i-1} = [x,z_i]$ for $i \neq p^j-1$ and $z_{p^j-2} = [x,z_{p^j-1}] - \gamma_j z_{p^m-2}$, $j = 0,\ldots,m-1$.

Then it is easy to see that

$$z_{-1} = ((\mathrm{ad}x)^{p^m-1} - \gamma_{m-1}(\mathrm{ad}x)^{p^{m-1}-1} - \ldots - \gamma_0)(z_{p^m-2}),$$

so $[x,z_{-1}] = 0$ and z_{-1} is a scalar multiple of x (Theorem 1.1 c), say $z_{-1} = \alpha x$. Now, putting $v_i = \alpha^{-1}z_i$ we get a basis $\{v_i : i = -1,\ldots,p^m-2\}$ of L with $v_{-1} = x$ and $[v_{-1},v_i] = v_{i-1}$ if $i \neq p^j-1$, $[v_{-1},v_{p^j-1}] = v_{p^j-2} + \gamma_j v_{p^m-2}$.

Using induction on $i+j$, with $i,j \geq 0$, and the Jacobi identity we see that $[v_{-1},[v_i,v_j]-\alpha_{ij}v_{i+j}]$ belongs to L_{p^m-2}. But for all $u \neq 0$ in S, $[v_{-1},u] \notin L_{p^m-2}$. Hence $[v_i,v_j] = \alpha_{ij}v_{i+j}$ for $i,j \geq 0$ and L is isomorphic to $W(1{:}m;\Gamma)$.

Remark: Let $A(1{:}m)$ be the divided power algebra in one indeterminate of dimension p^m. That is, $A(1{:}m)$ has a basis $\{x^i : i = 0,\ldots,p^m-1\}$ with $x^ix^j = \binom{i+j}{i}x^{i+j}$ and $x^{i+j} = 0$ if $i+j > p^m-1$. Let ∂ be the derivation of $A(1{:}m)$ given by $\partial(x^i) = x^{i-1}$ and δ_Γ the derivation given by $\delta_\Gamma(x^i) = 0$ if $i \neq p^j$, $\delta_\Gamma(x^{p^j}) = \gamma_j x^{p^m-1}$, where $\Gamma = (\gamma_0,\ldots,\gamma_{m-1})$. Finally, let $D_\Gamma = \partial + \delta_\Gamma$. Then the set $L(A(1{:}m),D_\Gamma)$ of the derivations of $A(1{:}m)$ of the form aD_Γ, with a in $A(1{:}m)$, is a Lie algebra (see [6]) with a basis $\{x^iD_\Gamma : i = 0,\ldots,p^m-1\}$. This basis has the same multiplication table as the standard basis of $W(1{:}m;\Gamma)$. Hence the algebras $W(1{:}m;\Gamma)$ are of the type studied by Ree in [6].

Notice that for $i > 0$ $x^iD_\Gamma = x^i\partial$.

2. Exponentials

Let $A = A(1{:}m)$ be the divided power algebra considered in the last remark. If $-1 \leq h,k \leq p^m-2$ then h and k can be uniquely expressed in the form $h = \sum_{i=0}^{m-1} h_ip^i$, $k = \sum_{i=0}^{m-1} k_ip^i$, where $0 \leq h_i,k_i \leq p-1$ for $i > 0$ and $-1 \leq h_0,k_0 \leq p-2$. We observe that $x^{h+1}\partial(x^{k+1}) = \binom{h+k+1}{h+1}x^{h+k+1}$. From [2; Lemma 2.3] we see that $\binom{h+k+1}{h+1} = 0$ if either:

a) $k_0 \neq -1$ and $h_0+k_0+1 \geq p$ or $h_i+k_i \geq p$ for some i.

b) $k_0 = -1$, $k_1 = \ldots = k_t = 0 \neq k_{t+1}$ and $h_0 \neq -1$ or some $h_i \neq 0$ $(i = 1,\ldots,t)$, or $h_{t+1}+k_{t+1}-1 \geq p$, or some $h_i+k_i \geq p$ $(i = t+2,\ldots,m-1)$.

Now, if $h \neq p^j-1$ and $h > 0$, it is easy to verify that $(x^{h+1}\partial)^p = 0$. We consider $\exp x^{h+1}\partial = \sum_{i=0}^{p-1} \frac{1}{i!}(x^{h+1}\partial)^i$.

<u>Lemma 2.1</u> For $h \neq p^t-1$, $h > 0$, $\exp x^{h+1}\partial$ is an automorphism of A.

<u>Proof:</u> $\exp x^{h+1}\partial - 1$ is nilpotent on A, so $\exp x^{h+1}\partial$ is nonsingular. Now

$$\exp x^{h+1}\partial(x^{j+1}) \exp x^{h+1}\partial(x^{k+1}) - \exp x^{h+1}\partial(x^{j+1}x^{k+1}) =$$
$$\sum_{s=p}^{2p-2} \sum_{t=s+1-p}^{p-1} \frac{1}{t!(s-t)!}(x^{h+1}\partial)^t(x^{j+1})(x^{h+1}\partial)^{s-t}(x^{k+1}).$$

If $j_0 = -1$ (and the same for $k_0 = -1$) and $h_0 \neq -1$ then $x^{h+1}\partial(x^{j+1}) = 0$. If $h_0 = -1$ then, since $h \neq p^t-1$, either $h_i > 1$ for some i, or $h_i = h_r = 1$ for some i and r with $i < r$. In both cases $(x^{h+1}\partial)^t(x^{j+1})(x^{h+1}\partial)^{s-t}(x^{k+1}) = 0$, where $s \geq p$. In the remaining cases we have $(x^{h+1}\partial)^t(x^{j+1})(x^{h+1}\partial)^{s-t}(x^{k+1}) = 0$ too. So $\exp x^{h+1}\partial$ is an automorphism.

Let δ_Γ, D_Γ and $L = L(A,D_\Gamma)$ be as in the last section. Then for $h > 0$, $h \neq p^j-1$, we have $(\operatorname{ad} x^{h+1}\partial)^p = 0$ in L (see [4]) and:

<u>Lemma 2.2</u> Let L and h be as above:

a) $\exp \operatorname{ad} x^{h+1}\partial(x^{j+1}\partial) = \exp x^{h+1}\partial \cdot x^{j+1}\partial \cdot \exp(-x^{h+1}\partial)$.

b) $\exp \operatorname{ad} x^{h+1}\partial(D_\Gamma) = \exp x^{h+1}\partial \cdot D_\Gamma \cdot \exp(-x^{h+1}\partial)$.

<u>Proof:</u> a) will be valid if

$$\sum_{s=p}^{2p-2} \sum_{t=s+1-p}^{p-1} (-1)^t \frac{1}{t!(s-t)!} (x^{h+1}\partial)^{s-t} \cdot x^{j+1}\partial \cdot (x^{h+1}\partial)^t = 0.$$

But, as in the proof of Lemma 2.1, from our list of instances in which $\binom{h+k+1}{h+1} = 0$ we see that $(x^{h+1}\partial)^{s-t} \cdot x^{j+1}\partial \cdot (x^{h+1}\partial)^t(x^{k+1}) = 0$.

For b) it is enough to use a) and to check that

$$\exp x^{h+1}\partial \cdot \delta_\Gamma \cdot \exp(-x^{h+1}\partial) = \delta_\Gamma \text{ and } \exp \operatorname{ad} x^{h+1}\partial(\delta_\Gamma) = \delta_\Gamma.$$

<u>Corollary 2.3</u> In $W(1{:}m;\Gamma)$, $\exp \operatorname{ad} \alpha v_i$ is an automorphism for every $i > 0$, $i \neq p^j-1$, and $\alpha \in F$. This automorphism is induced by an automorphism of $A(1{:}m)$.

<u>3. Isomorphisms</u>

The aim of this section is the study of conditions for $W(1{:}m;\Gamma)$ and $W(1{:}m;\Gamma')$ to be isomorphic. We begin with an easy observation derived from Theorem 1.3:

<u>Lemma 3.1</u> Let L and L′ be Lie algebras of dimension $p^m > 3$, with respective subalgebras of codimension one containing no nonzero ideal S and S′. Then L is isomorphic to L′ if and only if there are elements x in L−S and x′ in L′−S′ such that the minimal polynomials of $\operatorname{ad} x$ and $\operatorname{ad} x'$ coincide.

Corollary 3.2 W(1:m;Γ) is isomorphic to the Zassenhaus algebra W(1:m) if and only if $\gamma_i \in F^{p^i}$ for all i.

Proof: If W(1:m;Γ) is isomorphic to W(1:m) then Theorem 1.1 b) shows that $\gamma_i \in F^{p^i}$ for all i. Conversely, if $\gamma_i = \alpha_i^{p^i}$ for all i, then we can find an element x in W(1:m), not contained in the subalgebra of codimension one, such that the minimal polynomial of ad x is $X^{p^m} - \gamma_{m-1}X^{p^{m-1}} - \ldots - \gamma_0 X$ (Theorem 1.1 b)). So W(1:m) is isomorphic to W(1:m;Γ) by Lemma 3.1.

The next result has been proved by Dzhumadil'daev in [5] and Varea in [7], as mentioned in the Introduction. It is also a direct consequence of the last Corollary:

Corollary 3.3 The field F is perfect if and only if the Zassenhaus algebras of dimension p^m, with $m > 1$, are the only Lie algebras of dimension $> p$ with a subalgebra of codimension one containing no nonzero ideal.

Corollary 3.4 The Witt algebra W(1:1) is the unique Lie algebra of dimension p with a subalgebra of codimension one containing no nonzero ideal.

In order to study the isomorphisms between the W(1:m;Γ)'s over nonperfect fields, we are going to study the minimal polynomials of the elements which are not contained in the subalgebra of codimension one. In view of Corollary 2.3, let us denote by Aut*(W(1:m;Γ)) the group of automorphisms of W(1:m;Γ) generated by $\{\exp \operatorname{ad} \alpha v_i : \alpha \in F,\ i > 0,\ i \neq p^j - 1\}$. Then, as in [4; Lemma 1] we get:

Lemma 3.5 Let $x \in W(1{:}m;\Gamma) - S$, where S is the subalgebra of codimension one, spanned by $\{v_i : i = 0,\ldots,p^m-2\}$. Then there is an element $s = v_{-1} + \sum_{i=1}^m \alpha_i v_{p^i-2}$ and an automorphism $\psi \in$ Aut*(W(1:m;Γ)) such that $\psi(x) = \alpha s$, for some α in F.

If we denote a standard basis of W(1:m;Γ) and W(1:m;Γ′) by $\{v_i : i = -1,\ldots,p^m-2\}$ and $\{v_i' : i = -1,\ldots,p^m-2\}$ respectively, then Lemma 3.5 implies:

Corollary 3.6 W(1:m;Γ) is isomorphic to W(1:m;Γ′) if and only if there is an element $s = v_{-1} + \sum_{i=1}^m \alpha_i v_{p^i-2}$ in W(1:m;Γ) and μ in F such that the minimal polynomials of ad s and ad $\mu v_{-1}'$ are equal.

As in [3; Section 2], given $\alpha_1,\ldots,\alpha_m$ in F we define polynomials $h_k(X)$ for $0 \le k \le m$ as follows:

$$h_0(X) = X$$
$$h_k(X) = X^{p^k} - (\alpha_1 X)^{p^{k-1}} - \ldots - (\alpha_{k-1}X)^p - \alpha_k X \quad \text{for } k > 0.$$

Now, if $s = v_{-1} + \sum_{i=1}^{m} \alpha_i v_{p^i-2} \in W(1{:}m;\Gamma)$, the matrix of the linear transformation $\mathrm{ad}\, s$ is the same as the matrix of the corresponding element in $W(1{:}m)$ except for the last row, where the γ_i's appear. Thus expanding the characteristic polynomial of this matrix by its last row and taking into account Theorem 1.1 b) we get:

<u>Lemma 3.7</u> The minimal and characteristic polynomial of $\mathrm{ad}\, s$ is $h_m(X) - \sum_{i=0}^{m-1} \gamma_i h_i(X)$.

That is, the minimal polynomial of $\mathrm{ad}\, s$ is $X^{p^m} - \beta_{m-1}X^{p^{m-1}} - \ldots - \beta_0 X$ where the coefficients are given by $\beta_i = (\alpha_{m-i})^{p^i} - \gamma_{m-1}(\alpha_{m-1-i})^{p^i} - \ldots - \gamma_{i+1}(\alpha_1)^{p^i} + \gamma_i$.

Since the characteristic polynomial of $\mathrm{ad}\, \mu v_{-1}$ is $X^{p^m} - \mu^{p^m - p^{m-1}} \gamma_{m-1} X^{p^{m-1}} - \ldots - \mu^{p^m - 1} \gamma_0 X$, we obtain from Corollary 3.6 the following result:

<u>Proposition 3.8</u> The algebras $W(1{:}m;\Gamma)$ and $W(1{:}m;\Gamma')$ are isomorphic if and only if there are scalars $\alpha_1,\ldots,\alpha_m$ and μ in F, $\mu \neq 0$, such that for $i = m-1,\ldots,0$ we have $(\alpha_{m-i})^{p^i} - \gamma_{m-1}(\alpha_{m-1-i})^{p^i} - \ldots - \gamma_{i+1}(\alpha_1)^{p^i} + \gamma_i = \mu^{p^m - p^i}\gamma_i'$.

In fact, given $\alpha_1,\ldots,\alpha_{m-1}$ and μ, α_m is determined. So there are only $m-1$ conditions for $\alpha_1,\ldots,\alpha_{m-1}$ and μ.

For $i = 0$ we have the condition

$$\alpha_m - \gamma_{m-1}\alpha_{m-1} - \gamma_{m-2}\alpha_{m-2} - \ldots - \gamma_1\alpha_1 + \gamma_0 = \mu^{p^m - 1}\gamma_0',$$

so taking $\mu = 1$, $\alpha_1 = \ldots = \alpha_{m-1} = 0$ and $\alpha_m = -\gamma_0$ we get:

<u>Corollary 3.9</u> For all Γ, $W(1{:}m;\Gamma)$ is isomorphic to $W(1{:}m;\Gamma')$ with $\Gamma' = (0,\gamma_1,\ldots,\gamma_{m-1})$.

<u>Theorem 3.10</u>

a) If F is perfect or $m = 1$, $W(1{:}m;\Gamma)$ is isomorphic to $W(1{:}m)$ for all Γ.

b) If F is not perfect and $m > 1$ there are infinitely many nonisomorphic algebras $W(1{:}m;\Gamma)$.

<u>Proof:</u> Assertion a) is just a restatement of Corollaries 3.3 and 3.4.

b) Let $\gamma \in F - F^p$, for $\mu \in F$ let $\Gamma_\mu = (0, \gamma + \mu^p\gamma^2, 0, \ldots, 0)$ and $L_\mu = W(1{:}m;\Gamma_\mu)$. Now, if $\mu \neq \nu$ and L_μ were isomorphic to L_ν, then there would exist $\alpha_1,\ldots,\alpha_{m-1}$ and ξ in F, $\xi \neq 0$, such that:

$$(\alpha_{m-i})^{p^i} = 0 \quad \text{for} \quad i = m-1,\ldots,2 \,,$$
$$(\alpha_{m-1})^p + (\gamma + \mu^p\gamma^2) = \xi^{p^m-p}(\gamma + \nu^p\gamma^2) \,.$$

Hence L_μ is isomorphic to L_ν if and only if

$$(\mu^p - \xi^{p^m-p}\nu^p)\gamma^2 + (1 - \xi^{p^m-p})\gamma + (\alpha_{m-1})^p = 0$$

for some α_{m-1} and ξ in F, $\xi \neq 0$. But the scalars $1 - \xi^{p^m-p}$ and $\mu^p - \xi^{p^m-p}\nu^p$, contained in F^p, are not both 0, and the minimal polynomial of γ over F^p is $X^p - \gamma^p$. Hence such scalars α_{m-1} and ξ do not exist.

Since F is not perfect, it is infinite, so $\{W(1:m;\Gamma_\mu) : \mu \in F\}$ is an infinite family of nonisomorphic Lie algebras.

4. Automorphisms

In this section, we shall use the preceding results on the minimal polynomials of the elements not contained in the subalgebra of codimension one, to determine the automorphism group of the Lie algebras $W(1:m;\Gamma)$.

Lemma 4.1 If $W(1:m;\Gamma)$ is not isomorphic to $W(1:m)$, then there is no element $s = v_{-1} + \sum_{i=1}^{m} \alpha_i v_{p^i-2}$ in $W(1:m;\Gamma)$, with some $\alpha_i \neq 0$, such that the minimal polynomials of $\mathrm{ad}\, s$ and $\mathrm{ad}\, \xi v_{-1}$ coincide for any ξ in F, $\xi \neq 0$.

Proof: Let us suppose that these minimal polynomials coincide. Then the elements $\alpha_1,\ldots,\alpha_{m-1}$ should verify

$$(\alpha_{m-i})^{p^i} - \gamma_{m-1}(\alpha_{m-1-i})^{p^i} - \ldots - \gamma_{i+1}(\alpha_1)^{p^i} + \gamma_i = \xi^{p^m-p^i}\gamma_i.$$

If $\xi = 1$, the only possibility is $\alpha_1 = \ldots = \alpha_{m-1} = 0$, contradicting the assumptions. If $\xi \neq 1$, α_1 exists only if $\gamma_{m-1} \in F^{p^{m-1}}$ because $(\alpha_1)^{p^{m-1}} = (\xi^{p-1} - 1)^{p^{m-1}}\gamma_{m-1}$. In this case α_2 exists only if $\gamma_{m-2} \in F^{p^{m-2}}$ and so on. Hence $\gamma_i \in F^{p^i}$ for all i and $W(1:m;\Gamma)$ is isomorphic to $W(1:m)$.

Lemma 4.2 Let ψ be an automorphism of $W(1:m;\Gamma)$, with $p^m > 3$, such that $\psi(v_{-1}) = v_{-1}$. Then ψ is the identity.

Proof: The subalgebra S spanned by $\{v_i : i \geq 0\}$, being the only subalgebra of codimension one, is preserved by ψ, and so are all the subspaces L_i in the filtration. Hence $\psi(v_{p^m-2}) = \xi v_{p^m-2}$ for some ξ in F, $\xi \neq 0$. From the relation $[v_{-1},v_i] = v_{i-1} + \beta_i v_{p^m-2}$ ($\beta_i = 0$ except perhaps for $i = p^j-1$, some j), we get $[v_{-1},\psi(v_i)] = \psi(v_{i-1}) + \xi\beta_i v_{p^m-2}$. So we obtain $\psi(v_i) = \xi v_i$ for all i. Hence $\xi = 1$ and ψ is the identity.

Theorem 4.3 If $W(1:m;\Gamma)$ is not isomorphic to the Zassenhaus algebra, then $Aut(W(1:m;\Gamma)) = Aut^*(W(1:m;\Gamma))$.

Proof: Let ψ be an automorphism of $W(1:m;\Gamma)$. Then $\psi(v_{-1}) \notin S$. From Lemma 3.5, there is an automorphism $\phi \in Aut^*(W(1:m;\Gamma))$ such that $\phi\psi(v_{-1}) = \lambda s$ for some scalar λ in F, where $s = v_{-1} + \sum_{i=1}^{m} \alpha_i v_{p^i-2}$, with $\alpha_1,\ldots,\alpha_m$ in F. Hence $ad\, s$ and $ad\, \lambda^{-1} v_{-1}$ have the same minimal polynomial. Lemma 4.1 implies that $s = v_{-1}$, and it is clear that $ad\, v_{-1}$ and $ad\, \lambda^{-1} v_{-1}$ have the same minimal polynomial if and only if $\lambda = 1$. Hence $\phi\psi(v_{-1}) = v_{-1}$, so $\psi = \phi^{-1}$ is in $Aut^*(W(1:m;\Gamma))$.

Let us consider now the Zassenhaus algebra $W(1:m)$ with $p^m > 3$:

Theorem 4.4 The group $Aut(W(1:m))$ is the semidirect product of the normal subgroup $Aut^*(W(1:m))$ and a subgroup isomorphic to F^* (the multiplicative group of nonzero elements in F).

Proof: Let ψ be an automorphism of $W(1:m)$. As in the last Theorem there is an automorphism $\phi \in Aut^*(W(1:m))$ such that $\phi\psi(v_{-1}) = \lambda s$, where $s = v_{-1} + \sum_{i=1}^{m} \alpha_i v_{p^i-2}$. The minimal polynomial of $ad\, s$ is $h_m(X)$ and $ad\, v_{-1}$ is nilpotent, so that $\alpha_1 = \ldots = \alpha_m = 0$ and $\phi\psi(v_{-1}) = \lambda v_{-1}$. For λ in F^* let us define the linear transformation ϕ_λ by $\phi_\lambda(v_i) = \lambda^{-i} v_i$. It is clear from the multiplication table of $W(1:m)$ that ϕ_λ is an automorphism. Moreover, $\phi_{(\lambda^{-1})} \cdot \phi \cdot \psi\, (v_{-1}) = v_{-1}$, so $\psi = \phi^{-1} \cdot \phi_\lambda$. If T denotes the subgroup $\{\phi_\lambda : \lambda \in F^*\}$, T is isomorphic to F^*. The matrix of any element in $Aut^*(W(1:m))$ with respect to the standard basis is triangular with 1's in the main diagonal, so $Aut^*(W(1:m)) \cap T = 1$. Finally, $\phi_\lambda \cdot \exp ad\, \alpha v_i \cdot \phi_{(\lambda^{-1})} = \exp ad\, \alpha\lambda v_i$, so $Aut^*(W(1:m))$ is a normal subgroup and the Theorem follows.

Remarks

a) $Aut^*(W(1:m;\Gamma))$ is a nilpotent group, since it consists of triangular matrices with 1's on the main diagonal. Hence $Aut(W(1:m))$ is a solvable group. This had been proved by Ree ([6; Theorem 12.13]) for $p \geq 5$.

b) With some restrictions on the field, Ree proved that every automorphism of $W(n:m_1,\ldots,m_n)$ (notation as in [8]) was induced by an automorphism of the divided power algebra $A(n:m_1,\ldots,m_n)$. In the case $n = 1$ that we treat here, with the only restriction of $p > 2$, Lemma 2.2 shows that $\exp ad\, \alpha v_i$ $(i > 0, i \neq p^j-1)$ is induced by $\exp \alpha x^{i+1}\partial$ and it is easy to see that the automorphism ϕ_λ is induced by the automorphism Φ_λ of $A(1:m)$ given by $\Phi_\lambda(x^i) = \lambda^{-i}x^i$. Moreover, the subgroup B_1 which appears in [8; Theorem 2] coincides in $Aut(A(1:m))$ with the subgroup generated by $\{\exp \alpha x^i\partial : \alpha \in F, i > 0, \ i \neq p^j\}$.

REFERENCES

[1] R.K. Amayo, Quasiideals of Lie algebras II. Proc. London Math. Soc. 33 (1976), 37–64.

[2] G.M. Benkart, I.M. Isaacs and J.M. Osborn, Lie algebras with self–centralizing ad–nilpotent elements. J. Algebra 57 (1979), 279–309.

[3] G.M. Benkart, I.M. Isaacs and J.M. Osborn, Albert–Zassenhaus algebras and isomorphisms. J. Algebra 57 (1979), 310–338.

[4] G. Brown, Cartan subalgebras of Zassenhaus algebras. Canad. J. Math. 27 (1975), 1011–1021.

[5] A.S. Dzhumadil'daev, Simple Lie algebras with a subalgebra of codimension one. Russ. Math. Surv. 40 (1985), 215–216.

[6] R. Ree, On generalized Witt algebras. Trans. Amer. Math. Soc. 83 (1956), 510–546.

[7] V.R. Varea, On the existence of ad–nilpotent elements and simple Lie algebras with subalgebras of codimension one. To appear in Proc. Amer. Math. Soc.

[8] R.L. Wilson, Classification of generalized Witt algebras over algebraically closed fields. Trans. Amer. Math. Soc. 153 (1971), 191–210.

1980 Mathematics Subject Classification: 17B50

Departamento de Matemática Aplicada
E.T.S.I.I.
Universidad de Zaragoza
50015 Zaragoza, Spain

This paper is in final form and no version of it will appear elsewhere.

FORMS OF RESTRICTED SIMPLE LIE ALGEBRAS

Shirlei Serconek* and Robert Lee Wilson**

Introduction

The classification of the finite-dimensional simple Lie algebras over a field P of prime characteristic p is an open problem. The problem naturally divides into two parts: classification for algebraically closed fields (itself an open problem, though for restricted algebras of characteristic > 7 the result is known [BW-88]) and classification for arbitrary fields (using the algebraically closed classification). The second problem is usually attacked by methods known as "descent theory".

The basic definition of descent theory is that of extension of the base field. If L is an algebra over Φ and $P \supseteq \Phi$ then $L_P = L \otimes_\Phi P$ is an algebra over P. If L_P is simple for every $P \supseteq \Phi$ we say L is a <u>central simple algebra</u> over Φ. If M is an algebra over P and $M \cong L_P$ (as algebras over P) we say L is a Φ-<u>form</u> of M. If L is a finite-dimensional simple algebra over Φ, then [Jac-62] there is a finite field extension $\Gamma \supseteq \Phi$, called the centroid of L, such that L is a central simple algebra over Γ. Therefore any simple algebra over Φ is a central simple algebra over some finite extension of Φ and any central simple algebra is a form of a simple algebra over an algebraically closed field. Thus to determine all simple algebras over arbitrary fields it is sufficient to determine all forms of simple algebras over algebraically closed fields. Since the restricted simple Lie algebras over algebraically closed fields of characteristic p > 7 are of classical or Cartan type ([BW-88]) the determination of all restricted simple Lie algebras over arbitrary fields of characteristic p > 7 would follow from the determination of all forms of algebras of classical or Cartan type.

---

* Partially supported by CNPq, Brazil

** Partially supported by The Institute for Advanced Study and by National Science Foundation Grant DMS-8603151

The determination of forms of classical algebras is discussed in detail in Chapter IV of [Sel-67]. For a classical simple Lie algebra L of type A, B, C, D (except D_4) of characteristic $p > 3$ all Φ-forms of L are known (in the sense that they correspond to central simple associative algebras with involution over Φ [Jac-41]). Also the forms of F_4 correspond to central simple exceptional Jordan algebras over Φ [Tom-53], [Bar-63],and the forms of G_2 correspond to Cayley-Dickson algebras over Φ [Jac-39], [Bar-63]. Furthermore, if L is a classical algebra over $P \supseteq \Phi$ (where characteristic $\Phi > 3$) and M, N are two Φ-forms of L, then there is some finite separable extension $P \supseteq \Gamma \supseteq \Phi$ such that $M_\Gamma \cong N_\Gamma$ [Bar-66],[Sel-67].

The determination of all forms of the classical algebras D_4, E_6, E_7 and E_8 remains an open problem.

The determination of the forms of the restricted simple Lie algebras of Carta type (for characteristic > 3) has been completed. For the Jacobson-Witt algebra W_m this result has been known for some time [Jac-43], [AS-69]. For the remainin restricted simple algebras of Cartan type it is a recent result of the authors [SW pre]. We will describe (without proofs) this determination here. We will begin (§1 by recalling the definition of the restricted simple Lie algebras of Cartan type an defining some related algebras. In §2 we will state the classification of forms of th restricted simple Lie algebras of Cartan type (Theorem 3). For the Jacobson-Wit algebra W_m this result is due to Jacobson [Jac-43] in case P/Φ is Galois or purel inseparable of exponent one and to Allen and Sweedler [AS-69] for arbitrary P/Φ. I §3 we will outline Jacobson's proof of this result in the purely inseparable exponen one case and in §4 we will show how Jacobson's proof can be extended (by replacin the use of derivations by higher derivations) to a proof for general P/Φ. Finally, i §5 we will indicate how the argument of §4 can be extended to all restricted simpl algebras of Cartan type, giving a proof of Theorem 3.

We will assume throughout that $P \supseteq \Phi$ are fields of characteristic $p > 3$.

§1. Algebras of Cartan type

Let $P \supseteq \Phi$ be fields of prime characteristic p and let m be a positive integer. Let $\mathbf{a}$ denote the m-tuple $(a_1, ..., a_m) \in \Phi^m$. Define

$$B_{\Phi,m}(\mathbf{a}) = \Phi[y_1, ..., y_m]/(y_1^p - a_1, ..., y_m^p - a_m)$$

and $W_{\Phi,m}(\mathbf{a}) = Der B_{\Phi,m}(\mathbf{a})$. Write $B_{\Phi,m}$ for $B_{\Phi,m}(\mathbf{0})$ and $W_{\Phi,m}$ for $W_{\Phi,m}(\mathbf{0})$ where $\mathbf{0} = (0, ..., 0)$. When the base field is clear from context we will write B_m for $B_{\Phi,m}$ and W_m for $W_{\Phi,m}$.

Then W_m is a restricted simple Lie algebra of dimension mp^m. It is called a Jacobson-Witt algebra.

Let B be a commutative associative algebra over Φ. Let $\Omega(B)$ denote the algebra of differential forms on B, i.e., $\Omega(B)$ is the exterior algebra, over B, on $Hom_B(Der B, B)$. Define

$$d : B \to Hom_B(Der B, B)$$

by

$$da(D) = Da$$

for all $a \in B, D \in Der B$. Now $Der B$ acts on $Hom_B(Der B, B)$ by

$$D(adb) = (Da)(db) + ad(Db)$$

for $a, b \in B, D \in Der B$. This action can be extended to $\Omega(B)$ so that $D(\alpha + \beta) = D\alpha + D\beta$, $D(\alpha \wedge \beta) = (D\alpha) \wedge \beta + \alpha \wedge (D\beta)$, and $D(a\alpha) = (Da)\alpha + a(D\alpha)$ for $a \in B$ and $\alpha, \beta \in \Omega(B)$.

Suppose $P \supseteq \Phi$, A is a commutative associative algebra over P and B is a Φ-form of A. Then it is well-known (cf. [Mat-86]) that $\Omega(A) \cong \Omega(B) \otimes P$. We will identify a differential form $\tau \in \Omega(B)$ with $\tau \otimes 1 \in \Omega(A)$. If $\omega \in \Omega(A)$ and $\tau \in \Omega(B)$ we write $\omega \sim \tau$ if and only if $\omega = \rho\phi(\tau)$ for some $0 \neq \rho \in P$ and some $\phi \in Aut(A)$.

The following lemma, due to Jacobson, gives the forms of $B_{P,m}$. We write P^p for the subfield $\{\rho^p \mid \rho \in P\}$ of P.

Lemma 1 ([Jac-43]). Let $P \supseteq \Phi$ and

$$\mathbf{a} \in (\Phi \cap P^p)^m. \tag{1.1}$$

Then $B_{\Phi,m}(\mathbf{a})$ is a Φ-form of $B_{P,m}$. Conversely, every Φ-form of $B_{P,m}$ is isomorphic (over Φ) to $B_{\Phi,m}(\mathbf{a})$ for some $\mathbf{a}$ satisfying (1.1).

Define the following elements of $\Omega(B_{P,m})$:

$$\omega_S = dy_1 \wedge \ldots \wedge dy_m, m \geq 3;$$

$$\omega_H = \sum_{i=1}^{r} dy_i \wedge dy_{i+r}, m = 2r \geq 2;$$

$$\omega_K = dy_{2r+1} + \sum_{i=1}^{r} (y_{i+r}dy_i - y_i dy_{i+r}), m = 2r + 1 \geq 3.$$

Using these differential forms we define the following algebras over P:

$$S_m = \{D \in W_m \mid D\omega_S = 0\}, m \geq 3;$$

$$CS_m = \{D \in W_m \mid D\omega_S \in P\omega_S\}, m \geq 3;$$

$$H_m = \{D \in W_m \mid D\omega_H = 0\}, m = 2r \geq 2;$$

$$CH_m = \{D \in W_m \mid D\omega_H \in P\omega_H\}, m = 2r \geq 2;$$

$$K_m = \{D \in W_m \mid D\omega_K \in B_{P,m}\omega_K\}, m = 2r + 1 \geq 3.$$

Let $L^{(n)}$ denote the nth derived algebra of the Lie algebra L. Then the algebra $S_m^{(1)}, m \geq 3$, is a restricted simple Lie algebra of dimension $(m-1)(p^m - 1)$. The algebra $H_m^{(2)}, m = 2r \geq 2$, is a restricted simple Lie algebra of dimension $p^m - 2$. The algebra $K_m^{(1)}, m = 2r+1 \geq 3$, is a restricted simple Lie algebra of dimension p^m if $m + 3 \not\equiv 0 \pmod p$ and of dimension $p^m - 1$ if $m + 3 \equiv 0 \pmod p$. The algebras $W_m, m \geq 1$; $S_m^{(1)}, m \geq 3$; $H_m^{(2)}, m = 2r \geq 2$; $K_m^{(1)}, m = 2r + 1 \geq 3$ are the restricted simple Lie algebras of Cartan type ([KS-66]).

The following facts about derivation algebras are well-known (cf. [SF-88]):

$$Der(W_m) = ad(W_m), m \geq 1;$$

$$Der(S_m^{(1)}) = ad(CS_m), Der(CS_m) = ad(CS_m), m \geq 3;$$

$$Der(H_m^{(2)}) = ad(CH_m), Der(CH_m) = ad(CH_m), m = 2r \geq 2;$$

$$Der(K_m^{(1)}) = ad(K_m), Der(K_m) = ad(K_m), m = 2r + 1 \geq 3.$$

For $\mathbf{a} \in \Phi^m$ define the following algebras over Φ:

$$S_{\Phi,m}(\mathbf{a} : \tau) = \{D \in W_{\Phi,m}(\mathbf{a}) \mid D\tau = 0\}, m \geq 3, \tau \sim \omega_S;$$

$$H_{\Phi,m}(\mathbf{a} : \tau) = \{D \in W_{\Phi,m}(\mathbf{a}) \mid D\tau = 0\}, m = 2r \geq 2, \tau \sim \omega_H;$$

$$K_{\Phi,m}(\mathbf{a} : \tau) = \{D \in W_{\Phi,m}(\mathbf{a}) \mid D\tau \in B_{\Phi,m}(\mathbf{a})\tau\}, m = 2r + 1 \geq 3, \tau \sim \omega_K.$$

Then the algebras $S_{\Phi,m}(\mathbf{a} : \tau)^{(1)}, H_{\Phi,m}(\mathbf{a} : \tau)^{(2)}$ and $K_{\Phi,m}(\mathbf{a} : \tau)^{(1)}$ are simple.

The following result is obvious.

Lemma 2. Assume that $\mathbf{a}$ satisfies (1.1). Then $W_{\Phi,m}(\mathbf{a})$ is a Φ-form of $W_{P,m}$ and $X_{\Phi,m}(\mathbf{a} : \tau)^{(2)}$ is a Φ-form of $X_{P,m}^{(2)}$ (for $X = S, H,$ or K).

§2. Main Result

The following theorem, which is the converse of Lemma 2, is the main result on forms of restricted simple Lie algebras of Cartan type.

Theorem 3. Let $P \supseteq \Phi$ be fields of characteristic $p > 3$. Let M be a restricted simple Lie algebra over Φ. Then:

a) (Jacobson [Jac-43], Allen-Sweedler [AS-69]) If $M_P \cong W_{P,m}$ then $M \cong W_{\Phi,m}(\mathbf{a})$ for some $\mathbf{a}$ satisfying (1.1).

b) If $M_P \cong S_{P,m}^{(1)}$ (for $m \geq 3$) then $M \cong S_{\Phi,m}(\mathbf{a}, \tau)^{(1)}$ for some $\mathbf{a}$ satisfying (1.1) and some $\tau \sim \omega_S$.

(c) If $M_P \cong H^{(2)}_{P,m}$ (for $m = 2r \geq 2$) then $M \cong H_{\Phi,m}(\mathbf{a},\tau)^{(2)}$ for some $\mathbf{a}$ satisfying (1.1) and some $\tau \sim \omega_H$.

(d) If $M_P \cong K^{(1)}_{P,m}$ (for $m = 2r+1 \geq 3$) then $M \cong K_{\Phi,m}(\mathbf{a},\tau)^{(1)}$ for some $\mathbf{a}$ satisfying (1.1) and some $\tau \sim \omega_K$.

§3. Jacobson's proof

We will now give a sketch of Jacobson's proof ([Jac-43]) of Theorem 3(a) in the case that P/Φ is a purely inseparable field extension of exponent one.

Jacobson's argument is based on two facts. The first is that if $P \supseteq \Phi$ is purely inseparable of exponent one then there exists some $\mu \in Der_\Phi P$ such that

$$\Phi = \{x \in P \mid \mu x = 0\}. \tag{3.1}$$

The second is that

$$ad : W_{P,m} \rightarrow Der(W_{P,m}) \text{ is surjective.} \tag{3.2}$$

Now let L be a Φ-form of $W_{P,m} = Der(B_{P,m})$. Let V denote the Φ-subalgebra of $B_{P,m}$ generated by $y_1, \ldots, y_m$, so that $B_{P,m} = V \otimes_\Phi P$. Define

$$D : B_{P,m} \rightarrow B_{P,m}$$

by

$$D = I_V \otimes \mu$$

where μ is as in (3.1) and I_V denotes the identity map on V.

Call $G \in End_\Phi B_{P,m}$ a <u>differential transformation of</u> $B_{P,m}$ <u>with respect to</u> μ if

$$G(ab) = G(a)b + aG(b)$$

and

$$G(\rho a) = \mu(\rho)a + \rho G(a)$$

for all $\rho \in P, a, b \in B_{P,m}$. For such a G let $I(G)$ denote $\{a \in B_{P,m} \mid G(a) = 0\}$. Clearly $I(G)$ is a Φ-subalgebra of $B_{P,m}$. The following lemma (from [Jac-43]) will be crucial in the proof of the theorem.

Lemma 4. Let μ, G and L be as above. If [G,L] = (0), then $I(G)$ is a Φ-form of $B_{P,m}$.

Then D is a differential transformation of $B_{P,m}$ with respect to μ. Define $\overline{D} = adD \otimes 1$, an endomorphism of $W_{P,m} = L \otimes_\Phi P$. Then $\overline{D} \in Der_P W_{P,m}$ and so, by (3.2), there exists $\delta \in W_{P,m}$ such that $adD = ad\delta$. Set $E = D - \delta$. Then E is a differential transformation of $B_{P,m}$ with respect to μ and $[E, L] = (0)$. Thus by Lemma 4, $I(E)$ is a Φ-form of $B_{P,m}$. Clearly $L \subseteq Der_\Phi(I(E))$ and comparison of dimensions shows that equality holds. Then Lemma 1 gives the result.

§4. Forms of W_m over arbitrary extensions

Theorem 3(a) (determination of the forms of W_m over arbitrary extensions) was proved by Allen and Sweedler ([AS-69]) as a special case of a very general theory of forms they developed using Hopf algebraic techniques. (It has also been proved by Waterhouse ([Wat-71]) using group schemes.)

In fact, for the algebras W_m (and indeed for all the restricted simple Lie algebras of Cartan type) the general Allen-Sweedler theory is not necessary. In these cases a fairly straightforward adaptation of Jacobson's argument (as presented in the previous section) suffices. In this section we will show how Jacobson's argument can be adapted to prove the Allen-Sweedler result, Theorem 3(a).

We begin by noting that it is only necessary to consider purely inseparable extensions P/Φ. This is because of the following result of Jacobson.

Lemma 5 [Jac-43]. Let P/Φ be a Galois extension and let M be a Lie algebra over Φ such that $M_P \cong W_{P,m}$. Then $M \cong W_{\Phi,m}$.

In view of this lemma (and the finite-dimensionality of W_m) it is sufficient to prove Theorem 3(a) in the case that P/Φ is a finite purely inseparable extension To adapt Jacobson's argument to this case we need a generalization of (3.1). Such generalization, involving higher derivations, follows from results of Sweedler [Swe-68 and of Weisfeld [Wei-65]. We now state the necessary definitions.

Let A be an algebra over Φ. Let $t \in \mathbb{N} \cup \{\infty\}$. A sequence of Φ-linear map pings $D = (D_0, D_1, ..., D_t)$ (or $(D_0, D_1, ...)$ if $t = \infty$) from A into itself is called higher derivation of A (over Φ) of length t if $D_0 = I_A$ and

$$D_i(ab) = \sum_{j=0}^{i} (D_j a)(D_{i-j} b)$$

for $0 < i < t+1; a, b \in A$. The set of all Φ-higher derivations of A of length t will b denoted by $HDer^t_\Phi(A)$.

Let $D \in HDer^t_\Phi(A)$. Let $I(D)$ denote

$$\{a \in A \mid D_j a = 0, 1 \leq j < t+1\}.$$

If $A = P$ is a field extension of Φ and $\mu \in HDer^t_\Phi(P)$ then $I(\mu)$ is a subfield of containing Φ.

Following Sweedler ([Swe-68]) we say that the extension P/Φ is modular if is a tensor product of primitively generated extensions of Φ. Sweedler has show that for arbitraty P/Φ there is a unique minimal $\Psi \supseteq P$ such that Ψ/Φ is modula Furthermore, Ψ/Φ is a finite purely inseparable extension if and only if P/Φ is. Ψ called the modular closure of P over Φ. By a result of Weisfeld ([Wei-65]) if P/Φ a modular purely inseparable field extension with an exponent (in particular, if $P/$ is a modular finite purely inseparable field extension) then there exists some $t \in$ and some $\mu \in HDer^t_\Phi(P)$ such that

$$\Phi = I(\mu). \tag{4.1}$$

One can show that Theorem 3 for general purely inseparable extensions can b deduced from Theorem 3 for modular purely inseparable extensions. Thus we wi henceforth assume that P/Φ is modular.

We now continue with some definitions and notation needed to finish describing he adaptation of Jacobson's proof.

Composition of higher derivations is defined as follows. Let A and t be as above nd $D, D' \in HDer^t_\Phi(A)$. For $0 \leq i < t+1$ set

$$(DD')_i = \sum_{j=0}^{i} D_j D'_{i-j}.$$

'hen $DD' = ((DD')_0, (DD')_1, ...) \in HDer^t_\Phi(A)$. Composition of higher derivations associative. Note that $(DD')_1 = D_1 + D'_1$, so that composition of higher derivations eneralizes addition of derivations. The higher derivation $(I_A, 0, ...)$ (denoted 1) is he identity element for composition of higher derivations.

Now let $t \in \mathbf{N}$ and A be an algebra over P and $\mu \in HDer^t_\Phi(P)$. Define

$$HDer^t_\mu(A)$$

o be the set of elements $D \in HDer^t_\Phi(A)$ such that

$$D_i(\rho a) = \sum_{j=0}^{i} \mu_j(\rho) D_{i-j}(a)$$

or all $\rho \in P, a \in A, 0 \leq i \leq t$. Observe that $HDer^t_1(A) = HDer^t_P(A)$. It is asily verified that if $\mu, \delta \in HDer^t_\Phi(P)$ and $D \in HDer^t_\mu(A), D' \in HDer^t_\delta(A)$ then $DD' \in HDer^t_{\mu\delta}(A)$.

Note that if $D = (D_0, D_1) \in HDer^1_\Phi(B_{P,m})$ then $D \in HDer^1_\mu(B_{P,m})$ if and only D_1 is a differential transformation of $B_{P,m}$ with respect to μ (as defined in §3).

Let A and t be as above and $D = (D_0, D_1, ...) \in HDer^t_\Phi(A)$. Set

$$D^* = (D^*_0, D^*_1, ...)$$

'here $D^*_0 = I_A$ and D^*_i is defined recursively by

$$\sum_{j=0}^{i} D_j D^*_{i-j} = 0$$

for $0 < i \leq t$. Then $D^* \in HDer^t_\Phi(A)$ and $DD^* = D^*D = 1$. D^* is called the inverse of D. Clearly $D^{**} = D$ and if $D \in HDer^t_\mu(A)$ then $D^* \in HDer^t_{\mu^*}(A)$. Also $I(\mu) = I(\mu^*)$ and $(DD')^* = D'^*D^*$ for $D, D' \in HDer^t_\Phi(A)$.

We now define the adjoint mapping. Let $t \in \mathbf{N}$, A be an algebra over P and $D \in HDer^t_\mu(A)$. For $0 \leq i \leq t$ and $e \in End_P A$ define

$$(ad\ D)_i e = \sum_{j=0}^{i} D_j e D^*_{i-j}.$$

Then $(ad\ D)_i \in End_P(End_P A)$. Thus

$$ad\ D = ((ad\ D)_0, (ad\ D)_1, \ldots) \in HDer^t_P(End_P A).$$

In fact,

$$(ad\ HDer^t_\mu(A))Der_P A \subseteq Der_P A.$$

Note that $(ad\ D)_1 e = D_1 e - e D_1$ so this definition generalizes the usual notion of adjoint.

We may now state the analogue of (3.2) we need to extend Jacobson's argument. This is due to Allen and Sweedler [AS-69]. Let $t \in \mathbf{N}$. Then

$$ad : HDer^t_P(B_{P,m}) \to HDer^t_P(W_{P,m}) \text{ is surjective.} \tag{4.2}$$

We also need the following strengthened version of Lemma 4.

Lemma 6. Let P and Φ be as above and let $t \in \mathbf{N}$ and $\mu \in HDer^t_\Phi(P)$ satisfy (4.1). Let L be a subset of $W_{P,m}$ such that

$$\{D \in W_{P,m} \mid [D, L] = (0)\} = (0).$$

Let $E \in HDer^t_\mu(B_{P,m})$. If

$$(ad\ E)_i L = (0)$$

for all $i, 0 < i \leq t$, then $I(E)$ is a Φ-form of $B_{P,m}$.

The proof of this result ([SW-pre]) uses the Jacobson-Bourbaki correspondence and coalgebra techniques.

With this background we may now describe the generalization of Jacobson's argument. (Recall that we assume P/Φ is modular.) Let L be a Φ-form of $W_{P,m}$. We let $t \in \mathbf{N}$ and $\mu \in HDer^t_\Phi(P)$ be as in (4.1). As in §3 we let V denote the Φ-subalgebra of $B_{P,m}$ generated by $y_1, ..., y_m$, so that $B_{P,m} = V \otimes_\Phi P$. We then define

$$D_i : B_{P,m} \to B_{P,m}$$

for $0 \leq i \leq t$, by

$$D_i = I_V \otimes \mu_i.$$

Then $D = (D_0, ..., D_t) \in HDer^t_\mu(B_{P,m})$. Then define $\overline{D}_i = (ad\ D)_i \otimes 1$, an endomorphism of $W_{P,m} = L \otimes_\Phi P$. It is immediate that $\overline{D} = (\overline{D}_0, ..., \overline{D}_t) \in HDer^t_P(W_{P,m})$. Then by (4.2) there exists $\delta \in HDer^t_P(B_{P,m})$ such that $ad\ \delta = \overline{D}$. Set $E = D\delta^*$. Then $E \in HDer^t_\mu(B_{P,m})$ satisfies the hypotheses of Lemma 6, so $I(E)$ is a Φ-form of $B_{P,m}$. It is clear that $L \subseteq Der_\Phi(I(E))$ and comparison of dimensions shows that equality holds. Then Theorem 3(a) follows from Lemma 1.

§5. Extension to algebras of types S, H, K

Let M be a restricted simple Lie algebra of type S, H or K over P. The following lemma (due to Serconek [Ser-80], cf. [SW-pre]) reduces the problem of determining the Φ-forms of M for arbitrary field extensions P/Φ to determining the forms for modular purely inseparable extensions.

Lemma 7. Let M be a restricted simple Lie algebra of Cartan type over P. Assume P/Φ is a Galois extension and N_1, N_2 are Lie algebras over Φ such that $(N_1)_P \cong (N_2)_P \cong M$. Then $N_1 \cong N_2$.

This lemma is proved by showing that $H^1(Gal(P/\Phi), Aut\ M) = \{1\}$ and using the well-known correspondence (cf. [Jac-66]) between Φ-forms of M and elements of $H^1(Gal(P/\Phi), Aut\ M)$. The computation of $H^1(Gal(P/\Phi), Aut\ M)$ is done by using the natural filtration of M and showing that

$$H^1(Gal(P/\Phi), Aut\ M) \cong H^1(Gal(P/\Phi), Aut\ (M_0/M_1)).$$

As $Aut\ (M_0/M_1)$ is a general linear group or is the semi-direct product of a symplectic group and the diagonal subgroup the triviality of $H^1(Gal(P/\Phi), Aut\ M)$ follows from known results ([Ser-68]).

Now assume that P/Φ is a modular finite purely inseparable extension. The surjectivity of ad plays an important role in the determination of the Φ-forms of $W_{P,m}$. However, in general, the restricted simple Lie algebras of Cartan type have outer derivations. Thus some care is necessary to adapt the argument of the previous section to arbitrary restricted simple Lie algebras of Cartan type. The results on derivation algebras of restricted simple Lie algebras of Cartan type stated in §1 show that if M is any such algebra, then $(Der\ M)^{(2)} = M$. From this it is easy to see that the forms of M and the forms of $Der\ M$ are in one-to-one correspondence. Since (as is also noted in §1) all derivations of $Der\ M$ are inner, we may, if an analogue of (4.2) is available, apply the methods of §4 to the algebras $CS_{P,m}, CH_{P,m}$ and $K_{P,m}$ and deduce the conclusion of Theorem 3.

It remains to give an analogue of (4.2) for algebras of types CS, CH and K. To do this note that there is a natural action of the components D_j of a higher derivation $D \in HDer^t_\mu(B_{P,m})$ on the algebra $\Omega(B_{P,m})$ of differential forms. This action is given by

$$D_i\lambda = \sum_{j=0}^{i} D_j\lambda(ad\ D)^*_{i-j}$$

for all $\lambda \in Hom_{B_m}(Der\ B_m, B_m)$ extended to $\Omega(B_{P,m})$ by

$$D_i(\alpha \wedge \beta) = \sum_{j=0}^{i}(D_j\alpha) \wedge (D_{i-j}\beta)$$

and

$$D_i(a\alpha) = \sum_{j=0}^{i}(D_ja)(D_{i-j}\alpha)$$

for $a \in B_{P,m}, \alpha, \beta \in \Omega(B_{P,m})$.

The following lemma (from [SW-pre]) gives the required analogue of (4.2).

Lemma 8. Let $t \in \mathbf{N}$. Then:

(a) $$ad : \{D \in HDer_P^t(B_m) \mid D_j\omega_S \in P\omega_S \text{ for all } j, 0 \leq j \leq t\}$$
$$\rightarrow HDer_P^t(CS_m) \text{ is surjective;}$$

(b) $$ad : \{D \in HDer_P^t(B_m) \mid D_j\omega_H \in P\omega_H \text{ for all } j, 0 \leq j \leq t\}$$
$$\rightarrow HDer_P^t(CH_m) \text{ is surjective;}$$

(c) $$ad : \{D \in HDer_P^t(B_m) \mid D_j\omega_K \in B_{P,m}\omega_K \text{ for all } j, 0 \leq j \leq t\}$$
$$\rightarrow HDer_P^t(K_m) \text{ is surjective.}$$

Using this lemma to replace (4.2) the proof of Theorem 3 may be completed.

References

-69] H. Allen and M. Sweedler, A theory of linear descent based upon Hopf algebraic techniques, J. Algebra 12 (1969), 242-294.

-63] R. T. Barnes, On derivation algebras and Lie algebras of prime characteristic. Ph.D. thesis, Yale University, 1963.

-66] —, On splitting fields for certain Lie algebras of prime characteristic, Proc. Amer. Math. Soc. 17 (1966), 930-935.

-88] R. E. Block and R. L. Wilson, Classification of the restricted simple Lie algebras, J. Algebra 114 (1988), 115-259.

-39] N. Jacobson, Cayley numbers and simple Lie algebras of type G, Duke Math. J. 5 (1939), 775-783.

-41] —, Classes of restricted Lie algebras of characteristic p. I, Amer. J. Math. 63 (1941), 481-515.

-43] —, Classes of restricted Lie algebras of characteristic p. II, Duke Math. J. 10 (1943), 107-121.

[Jac-62] —, Lie algebras, Interscience Tracts in Pure and Applied Math. 10, Int- science, New York, 1962.

[Jac-66] —, Forms of algebras, Yeshiva Sci. Confs. 1 (1966), 41-71.

[KS-66] A. I. Kostrikin and I. R. Šafarevič, Cartan pseudogroups and Lie p-algeb (Russian), Dokl. Akad. Nauk SSSR 168(1966), 740-742; English transl., Sov Math. Dokl. 7 (1966), 715-718.

[Mat-86] H. Matsumura, Commutative ring theory, Cambridge Studies in Advanc Mathematics 8, Cambridge University Press, Cambridge, 1986.

[Sel-67] G. B. Seligman, Modular Lie algebras, Ergebnisse der Mathematik und ih Grenzgebiete 40, Springer-Verlag, New York, 1967.

[Ser-80] S. Serconek, Forms of restricted simple Lie algebras of Cartan type. Ph. thesis, Rutgers University, 1980.

[SW-pre] S. Serconek and R. L. Wilson, Classification of forms of restricted simple I algebras of Cartan type, Comm. Algebra, to appear.

[Ser-68] J. P. Serre, Corps Locaux, Hermann, Paris, 1968.

[SF-88] H. Strade and R. Farnsteiner, Modular Lie algebras and their represen tions, Marcel Dekker, New York, 1988.

[Swe-68] M. E. Sweedler, Structure of inseparable extensions, Annals of Math. 87 (196 401-410.

[Tom-53] M. L. Tomber, Lie algebras of type F, Proc. Amer. Math. Soc. 4 (195 759-768.

[Wat-71] W. C. Waterhouse, Automorphism schemes and forms of Witt Lie algebras, Algebra 17 (1971), 34-40.

[Wei-65] M. Weisfeld, Purely inseparable extensions and higher derivations, Tra Amer. Math. Soc. 116 (1965), 435-450.

1980 AMS subject classification (1985 revision) 17B50

Department of Mathematics
Rutgers University
New Brunswick, NJ 08903
USA

This paper is in final form and no version of it will be submitted elsewhere.

THE SUBALGEBRA LATTICE OF A SUPERSOLVABLE LIE ALGEBRA

Vicente R. Varea*

Introduction.

Let L be a finite dimensional Lie algebra over a field F and $\mathcal{L}(L)$ denote the lattice of subalgebras of L. We are concerned with the relationship between the structure of L and that of $\mathcal{L}(L)$. There are numerous results in the literature relating both structures in the case when F has characteristic zero. For instance, lattice-theoretical characterizations of solvable and supersolvable Lie algebras have been obtained, Lie algebras with special kinds of subalgebra lattices have been described and several concepts invariant under lattice isomorphisms have been studied (for example, see Amayo [1], Amayo-Schwarz [2], Barnes [4,5], Gein [8,9,10], Gein-Muhin [11], Goto [12], Kolman [14,15], Towers [19,20,21], Varea [22]). All these results show that the lattice $\mathcal{L}(L)$ determines to a great extent the structure of L when the ground field F has characteristic zero.

However, little seems to be known about the relationship between the structure of L and that of $\mathcal{L}(L)$ when F has prime characteristic. We recall the recent classification of the Lie algebras with modular lattices of subalgebras due to Lashi [16] and the also recent characterization of the Lie algebras with a subalgebra lattice of length 2 due to Premet [17,18]. Moreover, when F is algebraically closed and char(F) $\neq$ 2, Gein [9] has proved that L is supersolvable if and only if $\mathcal{L}(L)$ has a modular chain. This result does not hold (necessarily) over fields that are not algebraically closed even when they have characteristic 0, since the 3-dimensional non-split simple Lie algebras have a modular chain.

In this paper we consider the case when F is perfect of characteristic p $\neq$ 2,3. We classify the Lie algebras whose subalgebra lattices have a modular chain and we

---

* Supported in part by CAICYT PR- 84-0778
This paper is in final form, and no version of it will be submitted for publication elsewhere.

characterize the supersolvable Lie algebras of dimension greater than three by the structure of their subalgebra lattices. To prove this we need to determine the effect the modular atoms of $\mathcal{L}(L)$ have on the structure of L. To do that we need to determine the Lie algebras containing no abelian subalgebras of dimension greater than 1. This last result allows us to give an easy proof of the above Lashi's result and to obtain a characterization of the Lie algebras with modular lattices of subalgebras in terms of their 1-dimensional subalgebras.

Throughout we shall consider finite dimensional Lie algebras over a perfect field F of characteristic $p \neq 2, 3$.

1. Lie algebras with modular lattices of subalgebras.

Let x be an element of a Lie algebra L. We denote by $E_L(x)$ the Engel subalgebra of L relative to x; that is, $E_L(x)$ is the Fitting null-component of L relative to the linear transformation ad x.

In [23], we proved that a Lie algebra L contains no 2-dimensional subalgebras if and only if $E_L(x) = Fx$ for every $x \in L - (0)$. Now we determine the Lie algebras containing no 2-dimensional subalgebras.

Lemma 1.1. The non-split 3-dimensional simple are the only Lie algebras of dimension greater than 2 containing no 2-dimensional subalgebras.

Proof. Let L be a Lie algebra of dimension greater than 1 containing no 2-dimensional subalgebras. Let $x \in L\text{-}(0)$. If $N_L(Fx) \neq Fx$, then there exists $y \in L\text{-}Fx$ such that $[yx] = tx$ for some $t \in F$. So $Fx + Fy$ is a 2-dimensional subalgebra, which is a contradiction. Then we have $N_L(Fx) = Fx$. Therefore, Fx is a Cartan subalgebra of L and $E_L(x) = Fx$ for every $x \in L\text{-}(0)$. As every subalgebra of L containing an Engel subalgebra is self-normalizing (see [3]), it follows that L must be simple. Now consider the centroid Γ of L. Since Γx is an abelian subalgebra of L, we obtain that $\Gamma x = Fx$. This yields $\Gamma = F$, so L is central-simple.

Let Ω denote an algebraic closure of F. By the preceding paragraph, the Lie algebra $L_\Omega = L \otimes_F \Omega$ is simple over Ω. Let $x \in L\text{-}(0)$. We have $E_{L_\Omega}(x) = (E_L(x))_\Omega = \Omega x$, so Ωx is a Cartan subalgebra of L_Ω. Thus L_Ω is either isomorphic to $sl(2,\Omega)$ or an Albert-Zassenhaus algebra. Let $L_\Omega = \Omega x \oplus \Sigma (L_\Omega)_\alpha$ be the Cartan decomposition of L_Ω relative to Ωx. By [7], $\dim(L_\Omega)_\alpha = 1$ for every non-zero root α. This yields that ad x is a semisimple transformation on L for every $x \in L$. Then, by Premet [18, Corollary 2], we obtain that L_Ω is a classical Lie algebra. Therefore $L_\Omega \cong sl(2,\Omega)$. We conclude that L is 3-dimensional non-split simple. The proof is complete.

By using Lemma 1.1 we obtain an easy proof of Lashi's result [16] and a characterization of the Lie algebras with modular lattices of subalgebras in terms of their 1-dimensional subalgebras.

A Lie algebra L is called almost-abelian if $L = V + Fa$ where V is an abelian ideal of L and a acts as the identity map on V.

Theorem 1.2. For a Lie algebra L the following are equivalent:

(1) The lattice $\mathfrak{L}(L)$ is modular,

(2) Each subalgebra of L of dimension 1 is either self-normalizing or an ideal of L.

(3) L is either abelian, almost-abelian or 3-dimensional non-split simple.

Proof. (1) implies (2) follows from Lemma 1.5 of [2]. (2) implies (1) is clear. (2) implies (3). Suppose first that $0 \neq Z(L) \neq L$ and take $z \in Z(L)\text{-}(0)$. Since $Z(L) \neq L$, there exist elements $x,y \in L$ such that $[xy] \neq 0$. Then we have $Fx \neq Fz \neq Fy$. Since $z \in N_L(Fx) \cap N_L(Fy)$, it follows that $N_L(Fx) \neq Fx$ and $N_L(Fy) \neq Fy$. Then, by (2), we have that Fx and Fy are ideals of L. This yields $[xy] \in Fx \cap Fy = 0$ which is a contradiction. We conclude that either $Z(L) = 0$ or L is abelian.

Then suppose that $Z(L) = 0$ and that L is not 3-dimensional non-split simple. By Lemma 1.1, it follows that L must contain a subalgebra S of dimension 2. Take a 1-dimensional ideal Fx of S. We find, $S \leq N_L(Fx)$, so $N_L(Fx) \neq Fx$ .By (2), we have $Fx \lhd L$. This yields that $C_L(x)$ is an ideal of codimension 1 in L since $Z(L) = 0$. Write $L = C_L(x) + Fy$. On the other hand, we note that $C_L(x)$ also satisfies (2) since L does. As $Z(C_L(x)) \neq 0$, from the preceding paragraph it follows that $C_L(x)$ is abelian. Thus $C_L(x)$

$\leq N_L(Fc)$ for every $c \in C_L(x)$. By (2), we obtain that $Fc \lhd L$ for every $c \in C_L(x)$. This yields that each element of $C_L(x)$ is an eigenvector of ad y. Therefore, ad y is a scalar transformation of $C_L(x)$. We conclude that L is almost-abelian. This completes the proof.

2. Lie algebras whose subalgebra lattices have a modular atom.

In this section we study the effect the modular atoms of $\mathfrak{L}(L)$ have on the structure of L.

First we need to determine the Lie algebras containing no abelian subalgebras of dimension greater than 1. For this we use Lemma 1.1.

Proposition 2.1. The 3-dimensional simple are the only Lie algebras of dimension greater than 2 containing no abelian subalgebras of dimension > 1.

Proof. Let L be a Lie algebra of dimension greater than 2 which contains no abelian subalgebras of dimension > 1. Then $C_L(x) = Fx$ for every $x \in L$-(0). By Lemma 1.1, we may assume that L contains a 2-dimensional non-abelian subalgebra S. Take a basis a, b for S with product [ab] = a. We have, $Fa < S \leq E_L(a)$. Since a acts nilpotently on $E_L(a)$ and $C_L(a) = Fa$, from Theorems 2.8 and 7.1 of [6] it follows that $E_L(a)$ is either 2-dimensional non-abelian, isomorphic to sl(2), or a form of an Albert-Zassenhaus algebra. In the latter case, by taking the standard basis $y_{-1}, y_o, \ldots, y_{p^n-2}$ for $E_L(a)$, in notation of [6,p.289], we find $[y_1, y_{p^n-2}] = o$ which is a contradiction. We conclude that $E_L(a)$ is either 2-dimensional non-abelian or isomorphic to sl(2).

Now let Ω be an algebraic closure of F. Write $L_\Omega = L \otimes_F \Omega$. We consider the decomposition $L_\Omega = (E_L(a))_\Omega \oplus \Sigma(L_\Omega)_\alpha$ of L_Ω into the generalized subspaces relative to ad a. Since a = [ab], we have that ad a = [ad a, ad b]. Then each $(L_\Omega)_\alpha$ is invariant under ad b (see [13,p.40]). This yields that the transformation ad a has trace zero on $(L_\Omega)_\alpha$. Therefore, p divides $\dim (L_\Omega)_\alpha$ for every non-zero root α. We obtain that $\dim L = \dim E_L(a) + rp$ for some $r \geq 0$. We conclude that $\dim L \equiv 2$ or 3 (mod p).

On the other hand, since the Cartan subalgebras of $L\Omega$ are precisely the minimal Engel subalgebras of $L\Omega$ (see [3]), we have that $E_{L_\Omega}(a)$ contains a Cartan subalgebra of

L_Ω. This Cartan subalgebra of L_Ω must be 1-dimensional since $E_L(a)$ is either 2-dimensional non-abelian or isomorphic to sl(2). Therefore, L_Ω has rank one.

Next, assume that L is a counterexample of minimal dimension. Let N be a proper ideal of L. Suppose dim N > 1. Then by the minimality of dim L it follows that N is either 2-dimensional non-abelian or 3-dimensional simple. Thus every derivation of N is inner in any of the two cases. Let $x \in L\text{-}N$. We have that there exists $y \in N$ such that $\operatorname{ad} x|_N = \operatorname{ad}_N y$. We find $[xy] = 0$, which is a contradiction. Thus dim N = 1. Write N = Fc. We have, $C_L(c) \triangleleft L$ and $\dim L/C_L(c) \leq 1$. But since $C_L(c) = Fc$, we find dim $L \leq 2$ a contradiction again. Consequently, L is simple. Since $Fx \leq \Gamma x \leq C_L(x) = Fx$ where Γ denotes the centroid of L, it follows $\Gamma = F$. Therefore L is central-simple, so L_Ω is a rank one simple Lie algebra over Ω. By [7], L_Ω is either isomorphic to $sl(2,\Omega)$ or an Albert-Zassenhaus algebra. If $L_\Omega \cong sl(2,\Omega)$, then dim L = 3 which is a contradiction. Hence L is a form of an Albert-Zassenhaus algebra. In particular, dim L = p^n for some n. We find $p^n \equiv 2$ or 3 (mod p), which contradicts $p \neq 2,3$. This completes the proof.

Let X,Y be subalgebras of L. Then we denote by $X \vee Y$ the subalgebra of L generated by X and Y.

<u>Theorem 2.2.</u> Let L be a Lie algebra. Assume that A is a modular atom of $\mathfrak{L}(L)$. Then either $\mathfrak{L}(L)$ is modular or A is an ideal of L.

<u>Proof.</u> Suppose that A is not an ideal of L. If $A \vee Fx = A + Fx$ for every $x \in L$, then L is almost-abelian by [1]. So $\mathfrak{L}(L)$ is modular. Thus we may assume that $A \vee Fb \neq A + Fb$ for some $b \in L$. Write $T = A \vee Fb$. By Lemma 1.5 of [2], A is maximal in T. We claim that T is 3-dimensional non-split simple. To prove this, let U , W be subalgebras of T such that $0 \neq U \leq W \neq T$ and $U \neq A$. From the modularity of A, it follows $(A \vee U) \cap W = U \vee (A \cap W)$. So $W = U$. This yields that every proper subalgebra of T is 1-dimensional. Then, by Premet [17], $\operatorname{ad}_T x$ is semisimple for every $x \in T$. It follows that T is 3-dimensional non-split simple by [18].

Now assume that L is a minimal counterexample. Since $\mathfrak{L}(T)$ is modular, we have $T \neq L$. Let M be a subalgebra of L such that $T < M$. By Theorem 1.2, the lattice $\mathfrak{L}(M)$ cannot be modular. On the other hand, since A is self-normalizing in L ([2]), we have that A is not an ideal of M. Then , as A is modular in $\mathcal{L}(M)$, it follows $M = L$ by the minimality of $\dim L$. Therefore, T is a maximal subalgebra of L.

Next, we show that L is simple. Let N be a minimal ideal of L. We have $L = T + N$ and $T \cap N = 0$, so $L/N \cong T$. We consider the subalgebra $A + N$. Since $\dim A = 1$, we have $A + N \neq L$. As $N_L(A) = A$, we have $A \ntrianglelefteq (A + N)$. It follows that $A + N$ is almost-abelian by the minimality of $\dim L$ and Theorem 1.2. Write $A = Fa$. Then we have that N is abelian and ad a is scalar on N. Thus there exists $t \in F\text{-}(0)$ such that $[ax] = tx$ for every $x \in N$. We find, $[[ab]x] = [a[bx]] + [[ax]b] = t[bx] + t[xb] = 0$ for every $x \in N$. It follows that the subalgebra $F[ab] + N$ is abelian. Write $c = [ab]$. Since T contains no 2-dimensional subalgebras, we have $c \in T\text{-}A$ and $T = A \vee Fc$. Let C be a maximal subalgebra of $Fc + N$ with $Fc \leq C$. We find, $T = A \vee Fc \leq A \vee C \leq L$. So $A \vee C = T$ or L by the maximality of T. On the other hand, we have

$$(A \vee C) \cap (Fc + N) = C \vee (A \cap (Fc + N)) = C$$

by the modularity of A. If $A \vee C = L$, then we find $Fc + N = C$ which is a contradiction. Therefore, $A \vee C = T$. Then we have, $Fc = T \cap (Fc + N) = C$. This yields that Fc is maximal in $Fc + N$, so $\dim (Fc + N) = 2$. Thus, $\dim N = 1$ and hence $\dim L/C_L(N) \leq 1$. But since L/N is simple and $[AN] \neq 0$, we have $C_L(N) = N$. This yields $\dim L = 2$, which is a contradiction. Consequently , L is simple.

Furthermore, since $N_L(A) = A$ and $\dim A = 1$, we have that L is central-simple of rank one. Then, by [7], it follows that either $\dim L = 3$ or L is a form of an Albert-Zassenhaus algebra. So $\dim L = 3$ or p^n for some n. Since $T \neq L$, we conclude that $\dim L = p^n > 3$.

Now let S be a 2-dimensional abelian subalgebra of L. Since $N_L(A) = A$, we have $S \cap A = 0$. Assume $A \vee S \neq L$. Then $A \vee S$ is almost-abelian by the minimality of $\dim L$ and Theorem 1.2. This yields $S \leq [A \vee S, A \vee S]$ since S is abelian and $\dim S = 2$. Therefore, there exists $\alpha \in F\text{-}(0)$ such that $[ax] = \alpha x$ for every $x \in S$. Thus S_Ω is contained in the root space $(L_\Omega)_\alpha$ of L_Ω relative to the Cartan subalgebra Ωa, where Ω denotes an algebraic closure of F. As $\dim (L_\Omega)_\alpha = 1$ by [7], we have $\dim S = 1$ which is a contradiction. Therefore, $A \vee S = L$. Let K be a maximal subalgebra of L with $S \leq K$. From the modularity of A , it follows $(A \vee S) \cap K = S \vee (A \cap K)$ whence $K = S$. This yields that S is a Cartan subalgebra of L. Let $L = S \oplus V$ be the Fitting decomposition of L relative to ad S. Take a non-zero linear form f of L such that f(V) = 0. Define the alternating bilinear form $B_f: L \times L \longrightarrow F$ by means of $B_f(x,y) = f([xy])$ for every $x,y \in L$. We denote by L_f the radical of B_f. We find,

$$f([SL]) = f([SS]) + f([SV]) = f([SV]) \leq f(V) = 0$$

so $S \leq L_f$. Since $f([LL]) = f(L) \neq 0$, we have $L_f \neq L$. As L_f is a subalgebra of L, it follows $S = L_f$ by the maximality of S. This yields that dim L/S is even, so dim L is even too. However, we have seen that $\dim L = p^n > 3$. This contradiction shows that L contains no 2-dimensional abelian subalgebras. Then, by using Proposition 2.1, we obtain that dim L = 3 which is a contradiction. This completes the proof.

Corollary 2.3 Let L be a Lie algebra. Assume that A is a modular atom of $\mathfrak{L}(L)$ and that $A \ntrianglelefteq L$. Then L is either almost-abelian or 3-dimensional non-split simple.

Proof. It follows from Theorems 2.2 and 1.2.

Corollary 2.4. The 3-dimensional non-split are the only simple Lie algebras containing a modular atom.

Proof. It follows from Corollary 2.3.

3. Lie algebras whose subalgebra lattices have a modular chain.

A chain $0 = M_0 < M_1 < \ldots < M_r = \mathfrak{L}$ of a lattice $\mathfrak{L}$ is said to be a modular chain if

(1) M_i is a modular element of $\mathfrak{L}$ for every $i = 0, \ldots, r$,

(2) M_i is maximal in M_{i+1} for every $i = 0, \ldots, r-1$.

In this section we classify the Lie algebras whose subalgebra lattices have a modular chain. By using this result we are able to characterize the supersolvable Lie algebras of dimension >3 by the structure of their subalgebra lattices

Theorem 3.1 Let L be a Lie algebra. Then the lattice $\mathfrak{L}(L)$ has a modular chain if and only one of the following holds:

1) L is supersolvable,

2) L is 3-dimensional non-split simple,

3) $L = S \oplus T$ where S is a supersolvable Lie algebra and T is a 3-dimensional non-split simple Lie algebra.

Proof Let $0 = M_0 < M_1 < \ldots < M_r = L$ be a modular chain of $\mathfrak{L}(L)$. We shall argue by induction on dim L. First suppose $M_1 \ntriangleleft L$. Since M_0 is maximal in M_1, we have dim $M_1 = 1$. Then Theorem 2.2 applies and $\mathfrak{L}(L)$ is modular. Thus, by Theorem 1.2, we have that L is as in (1) or (2).

Now assume $M_1 \triangleleft L$. Then we consider the Lie algebra L/M_1. Clearly $0 = M_1/M_1 < \ldots < M_r/M_1 = L/M_1$ is a modular chain of $\mathfrak{L}(L/M_1)$. By the inductive hypothesis, L/M_1 is either as in (1), (2) or (3). If L/M_1 is supersolvable, then L is supersolvable too, so L is as in (2). Next suppose that L/M_1 is 3-dimensional non-split simple. Write $M_1 = Fx$. Then there exists a basis e_1, e_2, e_3, x for L with product

$$[e_1,e_2] = e_3+\lambda x,\ [e_2,e_3] = \alpha e_1 + \mu x,\ [e_3,e_1] = \beta e_1 + \gamma x$$

and $[e_i,x] = t_i x$ for $i=1, 2, 3$,

where $\lambda, \alpha, \mu, \beta, \gamma, t_i \in F$ and $\alpha\beta \neq 0$ (see [13], p. 13).

By the Jacobi identity, $[[e_1,e_2],x] + [[e_2,x],e_1] + [[x,e_1]e_2] = 0$; whence $[e_3,x] + t_2[x,e_1] + t_1[e_2,x] = t_3x = 0$ so $t_3=0$. Similarly, we find $t_1=t_2=0$. Thus, $[e_i,x]=0$ for $i=1,2,3$. This yields that $L^{(1)}$ is the span of $[e_1,e_2]$, $[e_2,e_3]$, and $[e_3,e_1]$. We see that $x \notin L^{(1)}$ and dim $L^{(1)} = 3$. Therefore, $L = Fx \oplus L^{(1)}$. Then $L^{(1)} \cong L/Fx$, so $L^{(1)}$ is 3-dimensional non-split simple. We conclude that L is as in (3).

The remaining situation is where L/M_1 is as in (3). Then we can write, $L/M_1 = S/M_1 \oplus T/M_1$ where S/M_1 is supersolvable and T/M_1 is 3-dimensional non-split simple. As in the preceding paragraph, we find $T=M_1 \oplus T^{(1)}$. Therefore, $L = S + T^{(1)}$ and $S \cap T^{(1)} = 0$. Since S/M_1 is supersolvable and dim $M_1=1$, we have that S is supersolvable. Since $T \triangleleft L$, it follows $T^{(1)} \triangleleft L$. We conclude that L is as in (3). This completes the proof in one direction.

The converse is clear.

Next, we consider the subalgebra lattice of a supersolvable Lie algebra. Let L, M be Lie algebras. Assume that L is supersolvable and that $\mathfrak{L}(M)$ is isomorphic to $\mathfrak{L}(L)$.Then, Gein proved in [9] that either M is supersolvable or $M/Z(M)$ is simple,

$\dim Z(M) \leq 1$ and $Z(M)$ is contained in every maximal subalgebra of M.

We will see that M is supersolvable or 3-dimensional non-split simple. Furthermore, we will obtain a lattice-theoretical characterization of the supersolvable Lie algebras.

Let L be a Lie algebra. A chain of r+1 subalgebras $S_0 > S_1 > \dots > S_r$ is said to have length r. The length of $\mathcal{L}(L)$ is defined to be the length of the longest chain of subalgebras of L.

Lemma 3.2. Let L be a Lie algebra such that the lattice $\mathcal{L}(L)$ has length 3. Then $\mathcal{L}(L)$ has just one modular atom if and only if one of the following holds:

1) L is 3-dimensional with basis {a,b,c} and multiplication [a,b]=0, [a,c]=0, [b,c]=a (L is the Heisenberg algebra).

2) L is 3-dimensional with basis {a,b,c} and multiplication [a,b]=0, [a,c]=a, [b,c] = a+b.

3) $L = A \oplus T$ where $\dim A = 1$ and T is a 3-dimensional non-split simple Lie algebra.

Proof. Suppose that $\mathcal{L}(L)$ has just one modular atom A. If $A \ntriangleleft L$, then by Theorem 2.2 it follows that $\mathcal{L}(L)$ is modular. So every atom of $\mathcal{L}(L)$ is modular which is a contradiction. Therefore, $A \triangleleft L$. Thus we can consider the Lie algebra L/A. We have $\mathcal{L}(L/A)$ has length 2. Then by using [18], we obtain that L/A is either 2-dimensional or 3-dimensional non-split simple. First suppose $\dim L/A = 2$. Then $\dim L = 3$. By inspection of the 3-dimensional Lie algebras (see [13,p.11]), we find that the algebras described in (1) and (2) are the only 3-dimensional Lie algebras having just one ideal of dimension 1. Hence, L is as in (1) or (2).

Next assume that L/A is 3-dimensional non-split simple. Then, as in the proof of Theorem 3.1, we have $L = A \oplus L^{(1)}$. Thus L is as in (3). This completes the proof in one direction.

To prove the converse, assume L is as in (1), (2) or (3). Clearly, $\mathcal{L}(L)$ is not modular. Then, from Theorem 2.2 it follows that each modular atom of $\mathcal{L}(L)$ must be an ideal of L. However, it is easy to check that L has a unique ideal of dimension 1. Consequently, L has just one modular atom. The proof is complete.

Remark. Assume that L is as in (1), or (2) in the above Lemma. Then L is supersolvable and the unique modular atom Fx is not complemented in $\mathfrak{L}(L)$.

However, every Lie algebra L described in (3) in the above Lemma is not supersolvable and the unique modular atom A is complemented in $\mathfrak{L}(L)$.

Lemma 3.3. (Gein-Muhin [11]). Let L be a Lie algebra satisfying the following conditions:

1) $\mathfrak{L}(L)$ has length 3.

2) $\mathfrak{L}(L)$ has just one modular atom A.

3) the modular atom A is complemented; that is, $A \vee S = L$ and $A \cap S = 0$.

Then, $L = A \oplus L^{(1)}$ and $L^{(1)}$ is 3-dimensional non-split simple.

Proof. It follows from Lemma 3.2 and the above Remark.

Definition 3.4. Let A be an atom of a lattice $\mathfrak{L}$ and let $A \le K \in \mathfrak{L}$. The interval $I = [0:K]$ is said to be an A-interval if A is modular in I and there exists $S \in I$ such that $K = A \vee S$ and $A \cap S = 0$.

Definition 3.5. A lattice $\mathfrak{L}$ is called a su-lattice if it satisfies the following conditions :

(1) $\mathfrak{L}$ has a modular chain.

(2) $\mathfrak{L}$ does not contain any A-interval of length 3.

Theorem 3.6. Let L be a Lie algebra. Then L is either supersolvable or 3-dimensional non-split simple if and only if $\mathfrak{L}(L)$ is a su-lattice.

Proof. First suppose that $\mathfrak{L}(L)$ is a su-lattice. Assume that L is neither supersolvable nor 3-dimensional non-split simple. Then, by using Theorem 3.1, we obtain that $L = S \oplus T$ where $S \neq 0$ is supersolvable and $T \neq 0$ is 3-dimensional non-split simple. Take a 1-dimensional ideal Fx of S. Put $K = Fx \oplus T$. Since $[0:K] = \mathfrak{L}(K)$, we find that the interval $[0:K]$ is an Fx-interval of length 3, which is a contradiction.

The converse follows from Theorem 3.1 and Lemma 3.3.

Corollary 3.7. Let L, M be Lie algebras. Assume that L is supersolvable and that $\mathcal{L}(M)$ is isomorphic to $\mathcal{L}(L)$. Then, M is either supersolvable or 3-dimensional non-split simple.

Proof . It follows from Theorem 3.6.

References

[1] Amayo, R. K.: Quasi-ideals of Lie algebras, Proc. London Math. Soc.33 (1976), 28-36.

[2] Amayo, R. K., Schwarz, J.: Modularity in Lie algebras, Hiroshima Math. J. 10 (1980), 311-322.

[3] Barnes, D. W.: On Cartan subalgebras of Lie algebras, Math. Z. 101 (1967), 350-355.

[4] Barnes, D. W.: Lattice isomorphisms of Lie algebras, J. Austral. Math. Soc. 4 (1964), 470-475.

[5] Barnes, D. W.: Lattice automorphisms of semisimple Lie algebras, J. Austral. Math. Soc. 16 (1973), 43-45.

[6] Benkart, G. M., Isaacs, I. M., Osborn, J. M.: Lie algebras with self-centralizing ad-nilpotent elements. J. Algebra 57 (1979), 279-309.

[7] Benkart, G. M., Osborn, J. M.: Rank one Lie algebras. Annals of Math. 119 (1984), 437-463.

[8] Gein, A.G.: Semimodular Lie algebras, Sibirsk. Mat. Z. 17 (1976), 243-248 (translated in Siberian Math. J. 17 (1976), 243-248.

[9] Gein, A. G.: Supersolvable Lie algebras and the Dedekind law in the lattice of subalgebras. Ural. Gos. Univ. Mat. Zap. 10 (1977), 33-42.

[10] Gein, A.G.: Modular subalgebras and projections of locally finite dimensional Lie algebras of characteristic zero, Ural. Gos. Univ. Mat. Zap. 13 (1983), 39-51.

[11] Gein, A. G., Muhin, Complements to subalgebras of Lie algebras, Ural. Gos. Univ. Mat. Zap. 12 (1980), 2, 24-48.

[12] Goto, M.: Lattices of subalgebras of real Lie algebras, J. Algebra 11 (1969), 6-24.

[13] Jacobson, N.: Lie Algebras, Wiley- Interscience, New York (1962).

[14] Kolman, B.: Semi-modular Lie algebras, J. Sci. Hiroshima Univ. Ser. A-1 29 (1965), 149-163.

[15] Kolman, B.: Relatively Complemented Lie algebras, J. Sci. Hiroshima Univ. Ser. A-1 31 (1967), 1-11.

[16] Lashi, A. A.: On Lie algebras with modular lattices of subalgebras, J. Algebra 99 (1986), 80-88.

[17] Premet, A. A.: Toroidal Cartan subalgebras of Lie p-algebras, and anisotropic Lie algebras of positive characteristic, Vestsi Akad. Navuk BSSR Ser. Fiz. Mat. Navuk,1 (1986), 9-14.

[18] Premet, A.A.: Lie algebras without strong degeneration, Math. USSR Sb. 57 (1987), 151-163.

[19] Towers, D. A.: Lattice isomorphisms of Lie algebras, Math. Proc. Phil. Soc. 89 (1981), 285-292 ; corrigenda, 95, (1984), 511-512.

[20] Towers, D. A.: Lattice automorphisms of Lie algebras, Arch. Math. 46 (1986), 39-43.

[21] Towers, D. A.: Semi-modular subalgebras of a Lie algebra, J. Algebra 103 (1986), 202-207.

[22] Varea, V. R.: Lie algebras whose maximal subalgebras are modular, Proc. of the Royal Soc. of Edinburgh, 94 A (1983), 9-13.

[23] Varea, V. R.: Lie algebras none of whose Engel subalgebras are in intermediate position, Comm. in Algebra, 15 (12), (1987), 2529-2543.

1980 Mathematics Subject Classification: 17B50

University of Zaragoza
50009 Zaragoza, (Spain)

LIE THEORETIC METHODS IN COHOMOLOGY THEORY

Rolf Farnsteiner

§0 Introduction

Since its inception in the early 1940's homological algebra has enjoyed a sustained vigorous development. This circumstance is partly contingent upon the unifying nature of the general theory, as expounded by Cartan and Eilenberg in their foundational book [2]. According to their approach the earlier cohomology theories of groups, associative algebras, and Lie algebras turned out to be manifestations of the general framework of algebraic complexes and derived functors. More specifically, all of the aforementioned theories could be subsumed under the cohomology theory of an augmented ring.

The applications of cohomology theory are manifold. Many classical structure theoretical results, for example, resort in their proofs to cohomological methods. Notably the theorems of Weyl, Levi-Malcev, and Maschke can be considered consequences of cohomology vanishing theorems (cf. [11]).

In this article, we present Lie theoretic methods which give rise to vanishing theorems in the abovementioned most general context. In contrast to the known approaches (cf. [3],[5]) our results only necessitate the existence of elements that are completely characterized by their actions on algebras and modules. Section 1 investigates the cohomology groups of an augmented ring and generalizes various results of [6] and [7]. The succeeding section refines this approach in a more special setting which is closely patterned after the situations encountered in the cohomology theories of groups and Lie algebras. We consider weight modules relative to certain subalgebras and establish conditions under which the cohomology groups with coefficients in the dual of a weight module coincide with those of its weight submodule. These techniques play an important role in the study of central extensions of Kac-Moody Lie algebras (cf. [8]). In addition, they give rise to a useful connection between the extension- and torsion functors of weight modules. Some specializations of the material presented here were recently employed by Kumar [19] in the determination of cohomology groups of symmetrizable Kac-Moody Lie algebras. By selecting a few illustrative examples we show in the concluding section how our approach harmonizes with the abovementioned concept of unification. Theorem 3.1, for instance, furnishes a generalization of results by Barnes [1] and Hochschild-Serre [12] who arrived at their respective theorems by utilizing two markedly different techniques.

All the main results of this paper possess counterparts concerning homology groups and torsion functors. Since their proofs only require minor modifications, we shall leave the details to the interested reader.

§1 A Vanishing Theorem for the Cohomology Groups of an Augmented Ring

Throughout this section, K is assumed to be a commutative ring, R a K-algebra. We denote by R^- the commutator algebra of R, i.e. the Lie algebra with underlying K-module R and Lie product

$$[r,s] := rs - sr \quad \forall\, r,s \in R.$$

The left multiplication in R^- effected by the element $u \in R$ is customarily designated by $\mathrm{ad}\,u$.

Definition(cf.[2]) : A triple (R,Q,τ) consisting of R, a left R-module Q, and a surjective module homomorphism $\tau : R \twoheadrightarrow Q$ is called an augmented ring.

An R-bimodule M is a K-module endowed with two operations (left and right) which, aside from satisfying the usual requisite properties, commute; i.e. $(r\cdot m)\cdot s = r\cdot(m\cdot s)$ $\forall\, r,s \in R$, $\forall\, m \in M$. We let $H^n(R,M)$ denote the n-th Hochschild cohomology group of R with coefficients in M (cf. [13] for definition). If N is a left R-module, then $\mathrm{Ext}^n_R\,(Q,N)$ is referred to as the n-th cohomology group of the augmented ring (R,Q,τ) with coefficients in N. A K-linear mapping $f : N \longrightarrow N$ is called locally nilpotent if $N = \bigcup_{n\geq 1} \ker f^n$.

Theorem 1.1. : Suppose that R and Q are K-projective and let N be a left R-module. If there is $u \in \ker\tau$ such that

(a) $\mathrm{ad}\,u : R^- \longrightarrow R^-$ is locally nilpotent

(b) u operates invertibly on N,

then $\mathrm{Ext}_R^n(Q,N) = (0) \quad \forall\ n \geq 0$.

Proof : Theorem 2.8a of [2,p.167] establishes the existence of natural isomorphisms $\mathrm{Ext}_R^n(Q,N) \cong H^n(R,\mathrm{Hom}_K(Q,N))$, where $\mathrm{Hom}_K(Q,N)$ has the canonical R-bimodule structure. Since $u \in \ker\tau$, we obtain

$$\tau((\mathrm{ad}\,u)(x)) = \tau(ux - xu) = u\tau(x) \qquad \forall\ x \in R.$$

Now let q be an element of Q and write $q = \tau(r)$. Owing to condition (a) there is an element $n \in \mathbb{N}$ such that $(\mathrm{ad}\,u)^n(r) = 0$. Hence

$$u^n q = u^n \tau(r) = \tau((\mathrm{ad}\,u)^n(r)) = 0,$$

proving that u operates locally nilpotently on Q. The theorem is now a direct consequence of [7,(4.10)]. $\square$

For convenience we shall henceforth exclusively consider algebras and modules over a fixed base field F. Let N be a left R-module with representation $\rho : R \longrightarrow \mathrm{End}_F(N)$. For any subset $S \subset R$ we define

$$N_o(S) := \bigcap_{s\in S} \bigcup_{n\geq 1} \ker \rho(s)^n.$$

The subspace $N_o(S)$ is called the Fitting-0-component of N relative to S. If $\mathrm{ad}\,s : R^- \longrightarrow R^-$ is locally nilpotent for every element $s \in S$, then $N_o(S)$ is an R-submodule of N (cf. [22,p.23]).

Theorem 1.2. : Let N be a left module for the augmented F-algebra (R,Q,τ). Suppose that $S \subset \ker\tau$ is a subset such that

(a) $\mathrm{ad}\,s : R^- \longrightarrow R^-$ is locally nilpotent for every $s \in S$

(b) $N/N_o(S)$ is finite dimensional.

Then there are isomorphisms $\mathrm{Ext}_R^n(Q,N) \cong \mathrm{Ext}_R^n(Q,N_o(S)) \quad \forall\ n \geq 0$.

Proof : We shall proceed by induction on $q := \dim_F N/N_o(S)$. Assuming $q > 0$, we obtain the existence of an element $u_o \in S$ such that $N_o(S) \subset \mathfrak{N} := N_o(u_o) \subset N$ and $N \neq \mathfrak{N}$. The definition of $N_o(u_o)$ entails in combination with (b) that u_o operates invertibly on $N/\mathfrak{N}$. By (1.1) the cohomology groups $\mathrm{Ext}_R^n(Q,N/\mathfrak{N})$ vanish and the long exact cohomology sequence

$$\longrightarrow \mathrm{Ext}_R^{n-1}(Q, N/\mathfrak{N}) \longrightarrow \mathrm{Ext}_R^n(Q,\mathfrak{N}) \longrightarrow \mathrm{Ext}_R^n(Q,N) \longrightarrow \mathrm{Ext}_R^n(Q, N/\mathfrak{N})$$

collapses into isomorphisms $\mathrm{Ext}_R^n(Q,\mathfrak{N}) \cong \mathrm{Ext}_R^n(Q,N)$ $\forall\, n \geq 0$. Since $\mathfrak{N}_o(S)$ and $N_o(S)$ coincide, we may now apply the inductive hypothesis in order to obtain isomorphisms

$$\mathrm{Ext}_R^n(Q,\mathfrak{N}) \cong \mathrm{Ext}_R^n(Q,N_o(S)).$$

The proof may now be concluded by combining the above isomorphisms. □

As mentioned in our introductory remarks, the notion of an augmented ring is sufficiently general to comprise the cohomology theories of groups, Lie algebras, restricted Lie algebras and associative algebras. We shall dwell briefly on the latter for the purpose of illustration. Let A be an associative F-algebra, with opposite algebra A^{op}, whose multiplication is defined via

$$a \cdot b = ba \quad \forall\, a,b \in A.$$

Let $A^e := A \otimes_F A^{op}$ denote the <u>enveloping algebra</u> of A. The left A^e-modules naturally correspond to the A-bimodules. Note that A is a left A^e-module with action defined by $(a\otimes b)\cdot x := axb$ $\forall\, a,b,x \in A$. The unique F-linear map $\tau : A^e \longrightarrow\!\!\!\!\rightarrow A$ which sends $a\otimes b$ onto ab is easily seen to be an augmentation. It was shown by Cartan and Eilenberg that the n-th Hochschild cohomology group of A with coefficients in a given A-bimodule M is isomorphic to $\mathrm{Ext}_{A^e}^n(A,M)$.

<u>Corollary 1.3.</u> : Let M be an A-bimodule and suppose that there is an element $u \in \ker\tau$ such that

(a) $\mathrm{ad}\,u : (A^e)^- \longrightarrow (A^e)^-$ is locally nilpotent

(b) u operates invertibly on M.

Then $H^n(A,M) = (0)$ $\forall\, n \geq 0$. □

<u>Remark</u>: Corollary 1.3 generalizes (3.3) and (4.4) of [7]. The latter result immediately follows from the observation that any given element $u \in A$ operating locally nilpotently on A via the adjoint representation gives rise to an element $v := u\otimes 1 - 1\otimes u \in \ker\tau$ which enjoys the same property. This reasoning also affords generalizations of various other results of [7].

<u>Supplemented algebras</u> constitute an important class of augmented rings.

Such an algebra is given by a pair (A,ε) consisting of an associative F-algebra A and an algebra homomorphism $\varepsilon : A \longrightarrow F$. Note that ε gives F the structure of a left A-module such that (A,F,ε) is an augmented algebra. For any left A-module M we put

$$H^n(A,M) := \mathrm{Ext}^n_A(F,M) \quad \forall\, n \geq 0.$$

The standard examples for supplemented algebras are universal enveloping algebras of Lie algebras as well as group algebras.

For future reference we record the succeeding direct consequence of theorem 1.2.

<u>Corollary 1.4.(cf.[7])</u> : Let (A,ε) be a supplemented algebra, $S \subset \ker\varepsilon$ a subset. Suppose that M is a left A-module such that

(a) $\mathrm{ad}\, s : A^- \longrightarrow A^-$ is locally nilpotent $\quad \forall\, s \in S$

(b) $M/M_0(S)$ is finite dimensional.

Then there are isomorphisms $H^n(A,M) \cong H^n(A,M_0(S)) \quad \forall\, n \geq 0.$ □

<u>§2 Cohomology Groups of Algebras with Involution</u>

Let A be an associative F-algebra. An <u>involution</u> $\omega : A \longrightarrow A$ is an antiautomorphism of order 2, that is an F-linear map of order 2 satisfying

$$\omega(ab) = \omega(b)\omega(a) \quad \forall\, a,b \in A.$$

Let M be a left A-module, then the dual space M^* obtains the structure of a left A-module by virtue of

$$(a\cdot f)(m) := f(\omega(a)\cdot m) \quad \forall\, a \in A,\ \forall\, m \in M.$$

Now suppose that $H \subset A$ is a Lie subalgebra of A^- such that there exists a subset R of Map(H,F), the set of mappings from H to F, with

(a) $A = \bigoplus_{\alpha\in R} A_\alpha(H)$ $\qquad$ (b) $H \subset A_0(H)$,

where $A_\alpha(H) := \bigcap_{h\in H} \bigcup_{n\geq 1} \ker(\mathrm{ad}\,h - \alpha(h)\,\mathrm{id}_A)^n$ denotes the "root space" of A with root α. The sub semigroup of Map(H,F) which is generated by R is called the <u>root lattice</u> of A and will be denoted by Q_R. Let $\rho : A \longrightarrow \mathrm{End}_F(M)$ be a representation of A. Following [9] we define for $\lambda \in$ Map(H,F) the "weight space"

$$M_\lambda(H) := \bigcap_{h\in H} \bigcup_{n\geq 1} \ker(\rho(h) - \lambda(h)\,\mathrm{id}_M)^n$$

and call M an <u>H-pseudoweight module</u> if there exists a subset $\Lambda \subset Map(H,F)$ such that $M = \bigoplus_{\lambda \in \Lambda} M_\lambda(H)$. If $\rho(h) - \lambda(h)\, id_M$ is nilpotent on $M_\lambda(H)$ for every $\lambda \in \Lambda$, then M will be referred to as an <u>H-weight module</u>. Given an H-pseudoweight module $M = \bigoplus_{\lambda \in \Lambda} M_\lambda(H)$, we define its <u>graded dual</u>

$$M^{\#} := \{ f \in M^* \; ; \; f(M_\lambda(H)) = (0) \quad \text{for all but finitely many } \lambda \in \Lambda \}.$$

<u>Lemma 2.1.</u> : Suppose that $\omega(H) \subset H$. Then the following statements hold :

(1) $\omega(A_\alpha(H)) \subset A_{-\alpha\circ\omega}(H) \quad \forall\, \alpha \in R$

(2) $M^{\#}$ is an A-submodule of M^*

(3) If M is an H-weight module, then $M^{\#}$ is an H-weight module with weight spaces $(M^{\#})_\xi(H) = \{ f \in M^* \; ; \; f(M_\lambda(H)) = (0) \text{ for } \lambda \neq \xi\circ\omega\}$. In particular, $(M^{\#})_\xi(H) \cong (M_{\xi\circ\omega}(H))^* \;\; \forall\, \xi \in Map(H,F)$ and $\Lambda^{\#} := \{ \lambda\circ\omega \; ; \; \lambda \in \Lambda \}$ is the set of weights of $M^{\#}$.

<u>Proof</u> : (1) Let x be an element of $A_\alpha(H)$ and put $y := \omega(x)$. Note that since ω is an involution we have for arbitrary $h \in H$

$$(ad\, h + \alpha(\omega(h))\, id_A)\circ \omega = -\, \omega\circ(ad\, \omega(h) - \alpha(\omega(h))\, id_A).$$

Given $h \in H$, there is n such that $(ad\, \omega(h) - \alpha(\omega(h))\, id_A)^n(x) = 0$. The preceding identity then yields

$$(ad\, h + \alpha(\omega(h))\, id_A)^n(y) = (-1)^n\, \omega((ad\, \omega(h) - \alpha(\omega(h))\, id_A)^n(x)) = 0.$$

This, however,qualifies y as an element of $A_{-\alpha\circ\omega}(H)$.

(2) Let $f \in M^{\#}$, $a \in A_\alpha(H)$, and $m \in M_\lambda(H)$. Owing to (1) we obtain $\omega(a)\cdot m \in M_{\lambda-\alpha\circ\omega}$. Now let $S_f := \{ \lambda \in \Lambda \; ; \; f(M_\lambda(H)) \neq 0 \}$. Then S_f is finite and the above shows that the bijection $\sigma : Map(H,F) \longrightarrow Map(H,F) \; ; \; \sigma(\lambda) = \lambda - \alpha\circ\omega$ takes $S_{(a\cdot f)}$ into S_f. Hence $a\cdot f \in M^{\#}$, proving that $M^{\#}$ is a submodule of M^*.

(3) We let $\rho^*: A \longrightarrow End_F(M^*)$ denote the contragredient representation of M given by $\rho^*(a)(f)(m) = f(\omega(a)\cdot m)$. For $\xi \in Map(H,F)$, let $N_\xi := \{ f \in M^* \; ; \; f(M_\lambda(H)) = (0) \text{ for } \lambda \neq \xi\circ\omega \}$. We shall show that N_ξ is contained in $(M^{\#})_\xi(H)$. Given $h \in H$, $f \in N_\xi$, and $m_\nu \in M_\nu(H)$ we obtain, observing

$$(\rho^*(h) - \xi(h)\, id_{M^*})(f) = f\circ(\rho(\omega(h)) - \xi(h)\, id_M),$$

$$(\rho^*(h) - \xi(h)\, id_{M^*})^n(f)(m_\nu) = f((\rho(\omega(h)) - \xi(h)\, id_M)^n(m_\nu))$$

$$= f((\rho(\omega(h)) - (\xi\circ\omega)(\omega(h))\, id_M)^n(m_\nu))$$

$\forall$ n $\geq$ 1. According to our present assumption, there is $n \in \mathbb{N}$ such that $(\rho(\omega(h))-(\xi\circ\omega)(\omega(h))\mathrm{id}_M)^n$ vanishes on $M_{\xi\circ\omega}(H)$. Since $\rho(\omega(h))-(\xi\circ\omega)(\omega(h))\mathrm{id}_M$ leaves $M_\nu(H)$ invariant, it follows that $(\rho^*(h) - \xi(h)\,\mathrm{id}_{M^*})^n(f)(m_\nu) = 0$ for $\nu \neq \xi\circ\omega$, while the case $\nu = \xi\circ\omega$ is obtained from the above computation. By definition of $M^\circledast$, we have $M^\circledast = \Sigma_{\xi\in\mathcal{X}} N_\xi = \Sigma_{\xi\in\mathcal{X}} (M^\circledast)_\xi(H)$, where $\mathcal{X} = \{ \lambda\circ\omega \ ; \ \lambda \in \Lambda \}$. According to results of [9] $M^\circledast$ is an H-pseudoweight module and $N_\xi = (M^\circledast)_\xi(H)$ $\forall\ \xi \in \mathcal{X}$. The latter identities also show that $M^\circledast$ is a weight module. □

Throughout the remainder of this section we shall assume that $\omega(H) \subset H$. For any two subsets $\Lambda,\Gamma \subset \mathrm{Map}(H,F)$ and $u \in H$, $\alpha \in F$ and $\lambda \in \Lambda$, we put

$$\Gamma(\lambda)^u_\alpha = \{ \gamma \in \Gamma \ ; \ \gamma(\omega(u))-\lambda(u) = \alpha \}.$$

<u>Theorem 2.2.</u> : Let $M = \bigoplus_{\lambda\in\Lambda} M_\lambda(H)$ and $N = \bigoplus_{\gamma\in\Gamma} N_\gamma(H)$ be two H-weight modules. Suppose there is $u \in H$ such that $\Gamma(\lambda)^u_{\alpha(u)}$ is finite $\forall\ \lambda \in \Lambda$ $\forall\ \alpha \in Q_R$. Then the canonical injection $N^\circledast \hookrightarrow N^*$ induces isomorphisms $\mathrm{Ext}^n_A(M,N^\circledast) \cong \mathrm{Ext}^n_A(M,N^*)$.

<u>Proof</u> : We shall first show that $\mathrm{Ext}^n_A(M,N^*/N^\circledast) = (0)$ $\forall$ $n\geq 0$. According to (2.9) of [9] it suffices to verify that the mapping $N^*/N^\circledast \longrightarrow N^*/N^\circledast$; $x \mapsto u\cdot x - (\alpha+\lambda)(u)x$ is invertible for every $\alpha \in Q_R$ and $\lambda \in \Lambda$.

Let ρ denote the representation associated to N. Given $\alpha \in Q_R$ and $\lambda \in \Lambda$, it readily follows from [9,(1.1)] that $\rho(\omega(u)) - (\alpha+\lambda)(u)\,\mathrm{id}_N$ is invertible on $N_\gamma(H)$ for every $\gamma \in \Gamma - \Gamma(\lambda)^u_{\alpha(u)}$. For $\gamma \in \Gamma$ we consider $\tau_\gamma : N_\gamma(H) \longrightarrow N_\gamma(H)$; $\tau_\gamma := ((\rho(\omega(u))-(\alpha+\lambda)(u)\,\mathrm{id}_N)|_{N_\gamma(H)})^{-1}$ for $\gamma \in \Gamma-\Gamma(\lambda)^u_{\alpha(u)}$ and $\tau_\gamma := 0$ otherwise. Now we define an operator

$$\Theta : N^* \longrightarrow N^* \text{ via } \quad \Theta(g)|_{N_\gamma(H)} := g\circ\tau_\gamma \quad \forall\ g \in N^* \quad \forall\ \gamma \in \Gamma.$$

Note that Θ leaves $N^\circledast$ invariant and thereby gives rise to a linear mapping $\Omega : N^*/N^\circledast \longrightarrow N^*/N^\circledast$. Let f be an element of N^*, $\gamma \in \Gamma-\Gamma(\lambda)^u_{\alpha(u)}$, $n_\gamma \in N_\gamma(H)$. As before, we let $\rho^* : A \longrightarrow \mathrm{End}_F(N^*)$ denote the contragredient representation of ρ. Then

$$(\rho^*(u) - (\alpha+\lambda)(u)\,\mathrm{id}_{N^*})(\Theta(f))(n_\gamma) = \Theta(f)(\rho(\omega(u)) - (\alpha+\lambda)(u)\,\mathrm{id}_N)(n_\gamma) =$$
$$= f(n_\gamma)\ f\circ\tau_\gamma\circ (\rho(\omega(u)) - (\alpha+\lambda)(u)\,\mathrm{id}_N|_{N_\gamma(H)})(n_\gamma)$$

while

$$\Theta((\rho^*(u) - (\alpha+\lambda)(u)\,id_{N^*})(f))(n_\gamma) = \Theta(f\circ(\rho(\omega(u)) - (\alpha+\lambda)(u)\,id_N))(n_\gamma)$$
$$= f\circ(\rho(\omega(u)) - (\alpha+\lambda)(u)\,id_N)\circ\tau_\gamma(n_\gamma) = f(n_\gamma).$$

Since $\Gamma(\lambda)^u_{\alpha(u)}$ is finite it follows that $(\rho^*(u) - (\alpha+\lambda)(u)\,id_{N^*})\circ\Theta(f) - f$ and $\Theta\circ(\rho^*(u) - (\alpha+\lambda)(u)\,id_{N^*})(f) - f$ are contained in $N^\circledast$ for every $f \in N^*$. Consequently, Ω is the inverse of the mapping $N^*/N^\circledast \longrightarrow N^*/N^\circledast$; $x \mapsto u\cdot x - (\alpha+\lambda)(u)x$. By (2.9) of [9] we now obtain $\mathrm{Ext}^n_A(M, N^*/N^\circledast) = (0)\ \forall\ n\geq 0$ and the long exact cohomology sequence associated to

$$(0) \longrightarrow N^\circledast \longrightarrow N^* \longrightarrow N^*/N^\circledast \longrightarrow (0)$$

yields the asserted isomorphisms. □

The above result specializes to extension functors of groups and Lie algebras by considering the group algebra and the universal enveloping algebra, respectively. The following Corollary, for instance, generalizes Lemma 2.6 of [19] which appears to be originally attributable to Duflo.

Corollary 2.3. : Let L be a Lie algebra over F with decomposition $L = \bigoplus_{\alpha\in R} L_\alpha(H)$, $H \subset L_0(H)$. Let $M = \bigoplus_{\lambda\in\Lambda} M_\lambda(H)$ and $N = \bigoplus_{\gamma\in\Gamma} N_\gamma(H)$ be two H-weight modules. Suppose there is $u \in H$ such that $\{ \gamma \in \Gamma ; (\gamma+\lambda)(u) = -\alpha(u) \}$ is finite for every $\alpha \in Q_R$ and $\lambda \in \Lambda$. Then the canonical injection $N^\circledast \hookrightarrow N^*$ induces isomorphisms

$$\mathrm{Ext}^n_{U(L)}(M,N^\circledast) \cong \mathrm{Ext}^n_{U(L)}(M,N^*).$$

Proof : Let $\omega : U(L) \longrightarrow U(L)$ denote the unique antiautomorphism of U(L) satisfying $\omega(x) = -x\quad \forall\ x \in L$. Then ω is an involution and the dual module structure defined by ω coincides with the U(L)-module structure induced by the contragredient representation. Since H is contained in L our present assumptions are equivalent to the finiteness of $\Gamma(\lambda)^u_{\alpha(u)}$. Hence (2.2) yields the asserted result. □

By setting M = F we retrieve the following result of [9] :

Corollary 2.4. : Let L be as in (2.3) and let $N = \bigoplus_{\gamma\in\Gamma} N_\gamma(H)$ be an H-weight module. Suppose there is $u \in H$ such that $\{ \gamma \in \Gamma ; \gamma(u) = -\alpha(u) \}$ is finite for every $\alpha \in Q_R$. Then the canonical injection $N^\circledast \hookrightarrow N^*$ induces isomorphisms $H^n(L,N^\circledast) \cong H^n(L,N^*)\ \ \forall\ n\geq 0$. □

Let G be a group with group algebra F[G]. The mapping $g \mapsto g^{-1}$ extends to an involution ω of F[G]. Hence every G-module M gives rise to a contragredient representation on M^* by defining

$$(g\cdot f)(m) := f(g^{-1}\cdot m) \quad \forall\, g \in G \;\; \forall\, f \in M^*.$$

The above results then yield, with the obvious modifications, isomorphisms for the cohomology of groups with coefficients in certain G-modules.

Returning to the general situation we shall investigate connections between the functors Ext and Tor. We require the following subsidiary result :

<u>Lemma 2.5.</u> : Let M be an H-weight module such that $M_\lambda(H)$ is finite dimensional for every $\lambda \in \Lambda$. Then M and $(M^*)^*$ are isomorphic.

<u>Proof</u> : We first note that $\Gamma : M \longrightarrow (M^*)^*$; $\Gamma(m)(f) = f(m)$ is a homomorphism of A-modules. Indeed, we have $\Gamma(a\cdot m)(f) = f(a\cdot m) = (\omega(a)\cdot f)(m) = \Gamma(m)(\omega(a)\cdot f) = (a\cdot\Gamma(m))(f)$. Now suppose that $m \in M_\lambda(H)$. Given $f \in (M^*)_\tau(H)$ it follows that $\Gamma(m)(f) = f(m) = 0$ unless $\lambda = \tau\circ\omega$, or equivalently $\tau = \lambda\circ\omega$. Consequently, $\Gamma(m) \in ((M^*)^*)_\lambda(H)$. Since the latter space is, in accordance with (2.1), canonically isomorphic with the bidual of $M_\lambda(H)$ and Γ is injective, we see that Γ maps M isomorphically onto $(M^*)^*$. □

Our next result provides an identification between the extension and torsion functors of A without leaving the subcategory of weight modules.

<u>Theorem 2.6.</u> : Let M and N be as in (2.2). Suppose that $N_\gamma(H)$ is finite dimensional for every $\gamma \in \Gamma$. If there exists an element $u \in H$ such that $\Omega(\lambda)^u_{\alpha(u)} := \{ \gamma \in \Gamma \,;\, \gamma(u) = (\lambda+\alpha)(u) \}$ is finite for all λ and α, then there exist isomorphisms

$$\mathrm{Ext}^n_A(M,N) \cong (\mathrm{Tor}^A_n(N^*,M))^* \quad \forall n \geq 0.$$

<u>Proof</u> : Let M,N be two arbitrary left A-modules. Then N obtains the structure of a right A-module by setting

$$n\cdot a := \omega(a)\cdot n \quad \forall\, n \in N.$$

Proposition 5.1 of [2,p.120] then provides isomorphisms

$$(*) \qquad \mathrm{Ext}^n_A(M,N^*) \cong (\mathrm{Tor}^A_n(N,M))^* \quad \forall n \geq 0.$$

It follows from (2.1) that the set of weights Γ^* of N^* coincides with $\{\gamma\circ\omega\ ;\ \gamma\in\Gamma\}$. The bijective mapping $f:\Gamma^*\longrightarrow\Gamma\ ;\ f(\gamma^*)=\gamma^*\circ\omega$ maps $\Gamma^*(\lambda)^u_{\alpha(u)}$ onto $\Omega(\lambda)^u_{\alpha(u)}$ for every λ and α. Consequently, $\Gamma^*(\lambda)^u_{\alpha(u)}$ is finite for all λ and α and a consecutive application of (2.5) and (2.2) in conjunction with (∗) now yields isomorphisms

$$\mathrm{Ext}^n_A(M,N)\cong\mathrm{Ext}^n_A(M,(N^*)^*)\cong\mathrm{Ext}^n_A(M,(N^*)^*)\cong(\mathrm{Tor}^A_n((N^*),M))^*\quad\forall n\geq 0.\qquad\square$$

<u>Definition</u> : Let N be a left A-module. A bilinear form $\{,\}:N\times N\longrightarrow F$ is called <u>contravariant relative to ω</u> if

$$\{a\cdot m,n\}=\{m,\omega(a)\cdot n\}\quad\forall\ m,n\in N\quad\forall\ a\in A.$$

<u>Corollary 2.7.</u> : Let N be a left A-module with a nondegenerate contravariant form $\{,\}$. Then the following statements hold :

(1) If $N=\oplus_{\gamma\in\Gamma}N_\gamma(H)$ is an H-weight module with finite dimensional weight spaces, then $\{,\}$ induces an isomorphism $N\cong N^*$.

(2) If, in addition, $\Gamma(\lambda)^u_{\alpha(u)}$ is finite $\forall\ (\alpha,\lambda)\in Q_R\times\Lambda$ for some $u\in H$, then there exist isomorphisms

$$\mathrm{Ext}^n_A(M,N)\cong(\mathrm{Tor}^A_n(N,M))^*\quad\forall n\geq 0.$$

(3) Suppose in addition to the requirements of (1) and (2) that M also possesses a nondegenerate contravariant form and that $M_\lambda(H)$ is finite dimensional $\forall\ \lambda\in\Lambda$. If there is $u\in H$ such that $\Lambda(\gamma)^u_{\alpha(u)}$ is finite $\forall\ (\alpha,\gamma)\in Q_R\times\Gamma$, then we have isomorphisms

$$\mathrm{Ext}^n_A(M,N)\cong\mathrm{Ext}^n_A(N,M)\quad\forall\ n\geq 0.$$

<u>Proof</u> : (1) We let $\rho:A\longrightarrow\mathrm{End}_F(N)$ denote the representation from A in N. We shall first show that $\{N_\gamma(H),N_\tau(H)\}=0$ whenever $\gamma\neq\tau\circ\omega$. Let $n\in N_\gamma(H)$, $m\in N_\tau(H)$. As $\gamma\neq\tau\circ\omega$, there is $h\in H$ such that $\gamma(h)\neq\tau(\omega(h))$. As n is contained in $N_\gamma(H)$ we can find an integer $k\geq 0$ such that $(\rho(h)-\gamma(h)\mathrm{id}_N)^k(n)=0$. On the other hand $\rho(\omega(h))-\gamma(h)\mathrm{id}_N$ is invertible on $N_\tau(H)$ proving the existence of $m'\in N_\tau(H)$ such that $(\rho(\omega(h))-\gamma(h)\mathrm{id}_N)^k(m')=m$. This gives rise to

$$\{m,n\}=\{(\rho(\omega(h))-\gamma(h)\mathrm{id}_N)^k(m'),n\}=\{m',(\rho(h)-\gamma(h)\mathrm{id}_N)^k(n)\}=0.$$

It follows that $\{,\}$ induces a homomorphism $\Theta:N\longrightarrow N^*$ of A-modules such that

$\Theta(N_\gamma(H)) \subset (N^*)_\gamma(H) \quad \forall \gamma \in \Gamma$. The nondegeneracy of $(,)$ implies the injectivity of Θ. In particular, we obtain $\gamma\circ\omega \in \Gamma$ as well as $\dim_F N_\gamma(H) \le \dim_F(N^*)_\gamma(H) = \dim_F N_{\gamma\circ\omega}$. Replacing γ by $\gamma\circ\omega$ yields equality throughout, whence $\Theta(N) = N^*$.

(2) We apply (1), (2.2), and Proposition 5.1 of [2, p.120] consecutively to obtain natural isomorphisms

$$\mathrm{Ext}^n_A(M,N) \cong \mathrm{Ext}^n_A(M,N^*) \cong \mathrm{Ext}^n_A(M,N^*) \cong (\mathrm{Tor}^A_n(N,M))^* \quad \forall n \ge 0.$$

(3) Our present assumptions yield isomomorphisms

$$\mathrm{Ext}^n_A(M,N) \cong (\mathrm{Tor}^A_n(N,M))^* \;, \quad \mathrm{Ext}^n_A(N,M)) \cong (\mathrm{Tor}^A_n(M,N))^* \;\forall n \ge 0.$$

According to [2,p.109] there are canonical isomorphisms

$$\mathrm{Tor}^A_n(N,M) \cong \mathrm{Tor}^{A^{op}}_n(M,N).$$

Since ω is an isomorphism from A into A^{op}, the latter groups are identical with $\mathrm{Tor}^A_n(M,N)$. The assertion can now be obtained by combining the above mappings.□

§3 Applications

In this section we shall illustrate how the preceding results may be employed in several classical and non-classical contexts. Cohomological methods play a central role in the proofs of various structure theorems in Lie and group theory. We begin with a generalization of a theorem by Hochschild-Serre concerning finite dimensional representations of reductive Lie algebras over fields of characteristic 0.

Theorem 3.1. : Suppose that F is a field of characteristic 0 and let L be a finite dimensional Lie algebra over F. Assume that $J,H \subset L$ are ideals such that

(a) $J = (0)$ or J is semisimple

(b) H is nilpotent

(c) $L = H \oplus J$.

Let M be a finite dimensional L-module such that $M^L := \{ m \in M \,;\, x\cdot m = 0 \;\; \forall x \in L \} = (0)$. Then $H^n(L,M) = (0) \quad \forall n \ge 0$.

Proof : Let U(L) denote the universal enveloping algebra of L, C(U(L)) its center, and $U(L)^+$ the kernel of the canonical supplementation map. We consider $\mathcal{B} := C(U(L)) \cap U(L)^+ + H$ and note that $\mathcal{B}$ is a sub Lie algebra of $U(L)^+$ which operates, owing to the nilpotency of H, locally nilpotently on U(L) via the adjoint representation. The module M decomposes into its Fitting components relative to $\mathfrak{B}$: $M = M_0(\mathfrak{B}) \oplus M_1(\mathfrak{B})$, with both summands being L-submodules of M. The general theory of algebraic complexes in conjunction with (1.4) yields isomorphisms

$$(*) \qquad H^n(L,M) \cong H^n(U(L),M) \cong H^n(U(L),M_0(\mathfrak{B})) \cong H^n(L,M_0(\mathfrak{B})).$$

The results of classical structure theory give rise to a decomposition $J = \oplus_{i=1}^r J_i$ $(r \geq 0)$, where the J_i are simple ideals of L. We let ρ_i denote the restriction of the representation $\rho : L \longrightarrow gl(M_0(\mathfrak{B}))$ to J_i. Consider the corresponding trace form κ_{ρ_i}. If κ_{ρ_i} is nondegenerate, then its Casimir element c_i is contained in $C(U(J_i)) \cap U(J_i)^+ \subset \mathfrak{B}$ and thereby operates nilpotently on $M_0(\mathfrak{B})$. We therefore obtain

$$0 = tr(\rho(c_i)) = \dim_F J_i.$$

As F has characteristic 0, the latter condition forces J_i to vanish. Consequently, κ_{ρ_i} is trivial for every $i \in \{1,\dots,r\}$ and Cartan's criterion (cf. [22,p.40]) ensures the solvability of $J_i/\ker \rho_i$. As J_i is simple, it follows that $J_i \cdot M_0(\mathfrak{B}) = (0)$. Hence J annihilates $M_0(\mathfrak{B})$. If $M_0(\mathfrak{B}) \neq (0)$, then the Engel-Jacobson theorem [22, p.16], applied to the H-module $M_0(\mathfrak{B})$, guarantees the existence of a nonzero element v of $M_0(\mathfrak{B})$ such that $h \cdot v = 0 \ \forall \ h \in H$. This, however, qualifies v as an element of $M_0(\mathfrak{B})^L \subset M^L = (0)$, a contradiction. Consequently, $M_0(\mathfrak{B}) = (0)$ and our result follows from (*). □

The arguments employed in the proof of the preceding result bear fruit even in nonclassical situations such as the cohomology theory of restricted Lie algebras. Let F be a field of positive characteristic $p > 0$. Suppose that (L,[p]) is a restricted Lie algebra with restricted universal enveloping algebra u(L) (cf. [22] for definitions). The trivial Lie homomorphism $L \longrightarrow F$ extends, as in case of ordinary Lie algebras, to a supplementation $\varepsilon : u(L) \longrightarrow F$. Following Hochschild [16], we define the restricted cohomology groups of (L,[p]) with coefficients in a given restricted L-module M by means of

$$H^n_*(L,M) := H^n(u(L),M).$$

Proposition 3.2. : Let $(L,[p])$ be a finite dimensional restricted Lie algebra, $\rho : L \longrightarrow gl(M)$ a finite dimensional, indecomposable restricted representation with nondegenerate trace form κ_ρ. Then $p \mid \dim_F L$ or $H_*^n(L,M) = (0) \quad \forall\, n \geq 0$.

Proof : We consider the subalgebra $\mathcal{B} := C(u(L)) \cap \ker\varepsilon$ and decompose M into its Fitting components relative to $\mathcal{B}$: $M = M_0 \oplus M_1$. As $\mathcal{B}$ lies centrally in u(L), both summands are stable under the action of L. The indecomposability of M in conjunction with (1.4) now entails the vanishing of $H_*^n(L,M)$ unless $M = M_0$. In that case, let $c \in U(L)$ denote the Casimir element of κ_ρ and consider its image $\hat{c}$ in u(L). By definition $\hat{c}$ belongs to $\mathcal{B}$ and we conclude

$$0 = tr(\rho(\hat{c})) = \dim_F L.$$

Consequently, p divides the dimension of L. □

Remark : The conclusion of (3.2) obviously retains its validity if the restricted cohomology groups are replaced by the ordinary ones. In that case the result is a direct consequence of the vanishing theorems of Chevalley-Eilenberg [3]. However, being strictly based on properties of Lie cocycles, their methods are intransferrable to the above context.

Example : Suppose that $p \geq 5$ and let F be algebraically closed. We shall show that the restricted cohomology groups $H_*^n(sl(2,F),V)$ of sl(2,F) with coefficients in a k-dimensional irreducible restricted sl(2,F)-module V vanish unless $k \in \{1,p-1,p\}$.

Let $\rho : sl(2,F) \longrightarrow gl(V)$ be the representation of V. The canonical basis of sl(2,F) will be denoted by $\{e,f,h\}$. It is well-known (cf. for instance p.208 of [22]) that there exists an element $v \in V$ such that $V = \bigoplus_{i=0}^{k-1} F\rho(f)^i(v)$ and $\rho(h)(\rho(f)^i(v)) = (k-1-2i)\rho(f)^i(v) \quad 0 \leq i \leq k-1 \leq p-1$. An elementary computation then reveals that

$$3\,tr(\rho(h)^2) = k(k-1)(k+1).$$

Since k is bounded by p, it follows that $\kappa_\rho(h,h) \neq 0$ unless $k \in \{1,p-1,p\}$. As sl(2,F) is simple, κ_ρ is nondegenerate in all but these three exceptional cases. The assertion now follows from (3.2).

In order to demonstrate the applicability of our results to infinite dimensional algebras we shall briefly consider the so-called contragredient Lie algebras. Suppose that F has characteristic 0 and let A be an $(n\times n)$-matrix of rank ℓ, $\mathfrak{g} = \mathfrak{g}(A)$ its contragredient Lie algebra. By definition $\mathfrak{g}$ admits a root space decomposition

$$\mathfrak{g} = \mathfrak{h} \oplus \bigoplus_{\alpha\in R} \mathfrak{g}_\alpha,$$

where $\mathfrak{h} = \mathfrak{g}_0$ is abelian and of dimension $2n-\ell$ and $\mathfrak{g}_\alpha = \{ x \in \mathfrak{g} ; [h,x] = \alpha(h)x \ \forall h \in \mathfrak{h} \}$ is finite dimensional for every $\alpha \in R$. There are linearly independent roots $\alpha_1,\dots,\alpha_n \in R$, called the simple roots of $\mathfrak{g}$, such that every root $\alpha \in R$ is of the form $\alpha = \sum_{i=1}^n m_i \alpha_i$, with all $m_i \in \mathbb{Z}$. We let Q_R designate the root lattice of $\mathfrak{g}$, which in this particular case coincides with the set of integral linear combinations of the simple roots. A nonzero element $\alpha = \sum_{i=1}^n m_i \alpha_i$ of Q_R is referred to as positive (>0) if $m_i \geq 0 \ \ 1\leq i\leq n$ and negative (<0) if $m_i \leq 0 \ \ 1\leq i\leq n$.

The algebra $\mathfrak{g}$ is generated by $\mathfrak{h}$ and $\{ e_1,\dots,e_n,f_1,\dots,f_n \}$, where $e_i \in \mathfrak{g}_{\alpha_i}$ and $f_i \in \mathfrak{g}_{-\alpha_i}$ $1\leq i\leq n$. The defining properties ensure the existence of a unique involution $\omega : U(\mathfrak{g}) \longrightarrow U(\mathfrak{g})$ satisfying $\omega(e_i) = f_i$, $\omega(f_i) = e_i$, and $\omega(h) = h \ \ \forall h \in \mathfrak{h}$.

We let $\mathfrak{n}^+$ and $\mathfrak{n}^-$ denote the subalgebras of $\mathfrak{g}$ which are generated by $\{ e_1,\dots,e_n \}$ and $\{ f_1,\dots,f_n \}$, respectively. Then $\mathfrak{g} = \mathfrak{n}^- \oplus \mathfrak{h} \oplus \mathfrak{n}^+$, and the Poincaré - Birkhoff - Witt theorem entails a decomposition

$$U(\mathfrak{g}) = U(\mathfrak{h}) \oplus (\mathfrak{n}^- U(\mathfrak{g}) + U(\mathfrak{g})\, \mathfrak{n}^+)$$

whose projection onto the first summand will be denoted by $\mathfrak{X}$. The bilinear form $(,) : U(\mathfrak{g})\times U(\mathfrak{g}) \longrightarrow U(\mathfrak{h})$;

$$(u_1,u_2) := \mathfrak{X}(\omega(u_1)u_2)$$

is customarily referred to as the Shapovalov bilinear form (cf. [19]).

A $\mathfrak{g}$-module M is said to belong to the category $\mathcal{O}$ if it is an $\mathfrak{h}$-weight module with finite dimensional weight spaces such that

(a) $\mathfrak{h}$ acts on M by semisimple transformations

(b) there exists a finite set $\mathfrak{F} \subset \mathfrak{h}^*$ such that the set Λ of weights of M is contained in $\{ \mu - \alpha ; \alpha \geq 0 , \mu \in \mathfrak{F} \}$.

A weight λ is called a highest weight if $x\cdot m = 0 \ \ \forall m \in M_\lambda , \forall x \in \mathfrak{n}^+$. If M is generated by a nonzero highest weight vector v_λ, then M is called a highest weight module. The irreducible objects of the Bernstein-Gel'fand-Gel'fand category

$\mathcal{O}$ are parametrized by linear forms of $\mathfrak{h}$. The irreducible $\mathfrak{g}$-module with highest weight λ is customarily denoted by $L(\lambda)$. The linear form λ extends to a homomorphism $U(\mathfrak{h}) \longrightarrow F$ of associative algebras, which we will also denote by λ.

Theorem 3.3.([17,Satz 1.6]) : Let $M = U(\mathfrak{g})v_\lambda$ be a highest weight module with highest weight λ. Then the following statements hold :

(1) $\mathcal{S}_\lambda : M \times M \longrightarrow F$; $\mathcal{S}_\lambda(u_1 \cdot v_\lambda, u_2 \cdot v_\lambda) = \lambda \circ \mathcal{S}(u_1,u_2)$ is a non-trivial, symmetric contravariant form on M.

(2) $\mathcal{S}_\lambda$ is nondegenerate if and only if M is irreducible.

Theorem 3.4. : Let $\lambda,\mu \in \mathfrak{h}^*$ and suppose that M is an object of $\mathcal{O}$. Then the following statements hold :

(1) $\mathrm{Ext}^n_{U(\mathfrak{g})}(M,L(\lambda)) \cong \mathrm{Tor}_n^{U(\mathfrak{g})}(L(\lambda),M)^* \quad \forall\, n \geq 0$, where $L(\lambda)$ has the structure of a right $U(\mathfrak{g})$-module given by $m \cdot u := \omega(u) \cdot m$.

(2) $\mathrm{Ext}^n_{U(\mathfrak{g})}(L(\lambda),L(\mu)) \cong \mathrm{Ext}^n_{U(\mathfrak{g})}(L(\mu),L(\lambda)) \quad \forall\, n \geq 0$.

Proof : Since ω leaves $\mathfrak{h}$ invariant the general assumptions of section 2 are valid for $U(\mathfrak{g})$ and the Lie subalgebra $\mathfrak{h} \subset U(\mathfrak{g})^-$. Let $h_1,\dots,h_n$ be elements of $\mathfrak{h}$ such that $\alpha_i(h_j) = \delta_{ij}$ $1 \leq i,j \leq n$ and consider $u := \sum_{i=1}^n h_i$. For $\alpha = \sum_{i=1}^n m_i \alpha_i \in Q_R$ we obtain $\alpha(u) = \sum_{i=1}^n m_i =: ht(\alpha)$. It follows that $Q_R^+(r) := \{ \beta \geq 0 \,;\, \beta(u) = r \}$ is finite for every element $r \in F$.

Let $M = \oplus_{\nu \in \Lambda} M_\nu$ be the weight space decomposition of M. If Γ denotes the set of weight of $L(\lambda)$, then $\Gamma \subset \{ \lambda - \alpha \,;\, \alpha \geq 0 \}$ and $\Gamma(\nu)^u_{\alpha(u)} = \{ \gamma \in \Gamma \,;\, \gamma(u) - \nu(u) = \alpha(u) \} \subset \{ \lambda - \beta \,;\, \beta \in Q_R^+((\lambda-\nu-\alpha)(u)) \} \quad \forall\, (\alpha,\nu) \in Q_R \times \Lambda$. Since the latter sets are according to our earlier observation finite the assertion now follows from a consecutive application of (3.3) and (2.7).

(2) Since, for $M = L(\mu)$, $\Lambda(\gamma)^u_{\alpha(u)}$ is contained in the finite set $\{ \mu - \beta \,;\, \beta \in Q_R^+((\mu-\gamma-\alpha)(u)) \}$ for $\alpha \in Q_R$ and $\gamma \in \Gamma$, (3) of (2.7) yields the desired result. $\square$

In conclusion we turn to the cohomology theory of groups and give a proof of a result on which Maschke's theorem is based :

Theorem 3.5. : Let G be a finite group whose order is not divisible by the characteristic of F. Suppose that M is a left G-module. Then $H^n(G,M) = (0) \ \forall\ n \geq 1$.

Proof : Let F[G] denote the group algebra of G, $\varepsilon : F[G] \longrightarrow F$ the trivial supplementation given by $\varepsilon(g) = 1 \ \ \forall\ g \in G$. The element $u := \mathrm{ord}(G)^{-1}\sum_{x\in G} x - 1$ belongs to $C(F[G]) \cap \ker \varepsilon$ and satisfies the equation $X(X+1) = 0$. The general theory of spaces with operators provides a decomposition $M = M_0 \oplus M_1$ of M into G-submodules, where

$$M_0 = \{\, m \in M \,;\, u\cdot m = 0 \,\}, \quad M_1 = \{\, m \in M \,;\, u\cdot m = -m \,\}.$$

Since the cohomology groups of G with coefficients in M are isomorphic to the corresponding cohomology groups of the augmented ring $(F[G],F,\varepsilon)$ it readily follows from (1.1) that $H^n(G,M_1) = (0) \ \ \forall\ n \geq 0$.

Let $c := \mathrm{ord}(G)^{-1}\sum_{x\in G} x$, then $c\cdot m = m \ \ \forall\ m \in M_0$. As $gc = c \ \ \forall\ g \in G$ we obtain $g\cdot m = m \ \ \forall\ m \in M_0$ proving that M_0 is a trivial G-module. The first cohomology group of G with coefficients in a trivial G-module is easily seen to vanish. Hence $H^1(G,M) = (0)$. As $\mathrm{Ext}^n_{F[G]}(F,M) \cong H^n(G,M)$ the latter statement shows that F is F[G]-projective. This, however, entails the vanishing of $H^n(G,M)$ for $n \geq 1$. □

References

[1] D.W. Barnes; On the Cohomology of Soluble Lie Algebras ; Math. Z. 101 (1967) 343-349.

[2] H. Cartan, S. Eilenberg; *Homological Algebra*, Princeton Univ. Press; Princeton, NJ, 1956.

[3] C. Chevalley, S. Eilenberg; Cohomology Theory of Lie Groups and Lie Algebras ; Trans. Amer. Math. Soc. 63 (1948) 85-124.

[4] J. Dixmier; Cohomologie des Algèbres de Lie nilpotentes ; Acta Sci. (Szeged) 16 (1955) 246-250.

[5] A.S. Dzhumadil'daev; Cohomology of Modular Lie Algebras ; Math. USSR-Sb. 47 (1984) 127-143.

[6] R. Farnsteiner; On the Cohomology of Associative Algebras and Lie Algebras ; Proc. Amer. Math. Soc. 99 (1987) 415-420.

[7] R. Farnsteiner; On the Vanishing of Homology and Cohomology Groups of Associative Algebras ; Trans. Amer. Math. Soc. 306 (1988) 651-665.

[8] R. Farnsteiner; Derivations and Central Extensions of Finitely Generated Graded Lie Algebras ; J. Algebra 118 (1988) 33-45.

[9] R. Farnsteiner; Cohomology Groups of Infinite Dimensional Algebras; Math. Z. (to appear).

[10] A. Guichardet; *Cohomologie des Groupes Topologiques et des Algèbres de Lie* Cedic/Fernand Nathan, Paris 1980.

[11] P.Hilton, U. Stammbach; *A Course in Homological Algebra*; Springer Graduate Texts vol. 4; New York, Heidelberg, Berlin 1971

[12] G.P. Hochschild, J.P. Serre; Cohomology of Lie Algebras ; Ann. of Math. 57 (1953) 591-603.

[13] G.P. Hochschild; On the Cohomology Groups of an Associative Algebra, Ann. of Math. 46 (1945) 58-67.

[14] G.P. Hochschild; On the Cohomology Theory for Associative Algebras ; Ann. of Math. 47 (1946) 568-579.

[15] G.P. Hochschild; Cohomology and Representations of Associative Algebras ; Duke J. Math. 14 (1947) 921-948.

[16] G.P. Hochschild; Cohomology of Restricted Lie Algebras ; Amer. J. Math. 76 (1954) 555-580

[17] J.C. Jantzen; *Moduln mit einem höchsten Gewicht*; Springer Lecture Notes in Mathematics Vol. 750 ; Berlin Heidelberg New York 1979.

[18] S. Kumar; A Homology Vanishing Theorem for Kac-Moody Algebras with Coefficients in the Category $\mathcal{O}$; J. Algebra 102 (1986) 444-462

[19] S. Kumar; Extension of the Category $\mathcal{O}g$ and a Vanishing Theorem for the Ext Functor for Kac-Moody Algebras ; J. Algebra 108 (1987) 472-490.

[20] C. Sen; The Homology of Kac-Moody Lie Algebras with Coefficients in a Generalized Verma Module ; J. Algebra 90 (1984) 10-17.

[21] C. Sen, G. Shen; Cohomology of Graded Lie Algebras of Cartan Type of Characteristic p ; Abh. Math. Sem. Univ. Hamburg 57 (1987) 139-156.

[22] H. Strade, R. Farnsteiner; *Modular Lie Algebras and their Representations*; Marcel Dekker Textbooks and Monographs 116, New York 1988.

[23] F.L. Williams; The Cohomology of Semisimple Lie Algebras with Coefficients in a Verma Module ; Trans. Amer. Math. Soc. 240 (1978) 115-127.

AMS Subject Classification : 16 A 61, 17 B 56, 17B 67, 18 G 15

Department of Mathematics
University of Wisconsin
Milwaukee, WI 53201

This paper is in final form and no version of it will appear elsewhere.

AN INTRODUCTION TO SCHUBERT SUBMODULES*

Mark E. Hall

0. Introduction

Let $\mathfrak{g}$ be an arbitrary (not necessarily symmetrizable) Kac-Moody algebra, with $\mathfrak{h}$ as its Cartan subalgebra, $\mathfrak{n}^+ = \oplus\sum_{\alpha>0} \mathfrak{g}_\alpha$ the sum of its positive root spaces, $\mathfrak{b} = \mathfrak{h} \oplus \mathfrak{n}^+$ a Borel subalgebra, and $\mathfrak{W}$ as its Weyl group. Let (V,π) be an integrable highest weight representation of $\mathfrak{g}$, with highest weight λ. For each $w \in \mathfrak{W}$, choose a nonzero element v_w in the one-dimensional weight space $V_{w(\lambda)}$, and let V_w be the $\mathfrak{b}$-module generated by v_w. These $\mathfrak{b}$-modules V_w are the Schubert submodules of V.

Schubert submodules have already begun to show their usefulness in the study of the representations of Kac-Moody algebras. For example, they appear in [J] and [Ku], in connection with the Demazure character formula, and in [H] and [LMNP], in connection with Verma bases, and their uses seem likely to continue to expand, if only because they are finite-dimensional, whereas most integrable highest weight representations of $\mathfrak{g}$ are infinite-dimensional.

Although Schubert submodules are beginning to be used by several researchers, there appears to be no organized exposition of them in the literature, and it is the purpose of this paper to fill that gap. Specifically, we will present here a precise definition of Schubert submodules and proofs of some of their basic properties. These properties include their finite-dimensionality, the fact that $V_{r_i w}$ is the $\mathfrak{g}_{(i)}$-submodule of V generated by V_w whenever $\ell(r_i w) > \ell(w)$, where $\mathfrak{g}_{(i)} = \mathbb{C}e_i \oplus \mathbb{C}h_i \oplus \mathbb{C}f_i$, a copy of $\mathfrak{sl}(2,\mathbb{C})$ in $\mathfrak{g}$, and the fact that V is the union of its Schubert submodules. There are, of course, other properties that we will not mention, but most of these can easily be obtained using the methods of this paper.

* Some of the results in this paper are contained in the author's doctoral thesis, written under the direction of G. M. Benkart at the University of Wisconsin-Madison in 1987.

Many thanks go to Robert Moody, whose work on Verma bases and Schubert submodules directly inspired my own, and to Georgia Benkart, who brought his ideas to my attention, and generally provided direction and encouragement throughout my graduate career.

1. Notation and Background

For the most part we will be following the notation of [K] in this paper. However, there are some differences, so we will use this section to carefully spell out our notation, and to remind the reader of a few basic facts about Kac-Moody algebras.

We will be working over a field $\mathbb{C}$ which is algebraically closed of characteristic zero. Let $A = (a_{ij})_{i,j=1}^{\ell}$ be a generalized Cartan matrix. That is, A is an integer matrix such that $a_{ii} = 2$ for all i, $a_{ij} \leq 0$ if $i \neq j$, and $a_{ij} = 0 \Rightarrow a_{ji} = 0$. Choose a triple $(\mathfrak{h},\Pi,\Pi^\vee)$, unique up to isomorphism, where $\mathfrak{h}$ is a vector space over $\mathbb{C}$ of dimension $2\ell - \operatorname{rank} A$, and $\Pi = \{\alpha_1,\ldots,\alpha_\ell\} \subseteq \mathfrak{h}^*$ and $\Pi^\vee = \{h_1,\ldots,h_\ell\} \subseteq \mathfrak{h}$ are linearly independent sets satisfying $\alpha_j(h_i) = a_{ij}$.

The Kac-Moody algebra $\mathfrak{g} = \mathfrak{g}(A)$ is the unique (up to isomorphism) Lie algebra over $\mathbb{C}$ which is generated by $\mathfrak{h}$ and the elements $e_1,\ldots,e_\ell$, $f_1,\ldots,f_\ell$; satisfies the relations

$$[\mathfrak{h},\mathfrak{h}] = (0),$$
$$[e_i,f_j] = \delta_{ij}h_i,$$
$$[h,e_i] = \alpha_i(h)e_i \quad \text{and} \quad [h,f_i] = -\alpha_i(h)f_i \quad \text{for all } h \in \mathfrak{h};$$

and has no nonzero ideals which intersect $\mathfrak{h}$ trivially. The abelian subalgebra $\mathfrak{h}$ is called the <u>Cartan subalgebra</u>. Let $\mathfrak{n}^+$ be the subalgebra generated by $e_1,\ldots,e_\ell$, $\mathfrak{n}^-$ the subalgebra generated by $f_1,\ldots,f_\ell$. We have the decomposition $\mathfrak{g} = \mathfrak{n}^- \oplus \mathfrak{h} \oplus \mathfrak{n}^+$. Set $\mathfrak{b} = \mathfrak{h} \oplus \mathfrak{n}^+$, and for each $i \in \{1,\ldots,\ell\}$, let $\mathfrak{g}_{(i)} = \mathbb{C}e_i \oplus \mathbb{C}h_i \oplus \mathbb{C}f_i$, a copy of $\mathfrak{sl}(2,\mathbb{C})$ in $\mathfrak{g}$.

Put $Q = \sum_{i=1}^{\ell} \mathbb{Z}\alpha_i$ and $Q^+ = \sum_{i=1}^{\ell} \mathbb{Z}^{\geq 0}\alpha_i$ (where $\mathbb{Z}^{\geq 0} = \{0,1,\ldots\}$), and for each $\alpha \in \mathfrak{h}^*$ let $\mathfrak{g}_\alpha$ denote the <u>root space</u> $\{x \in \mathfrak{g} \mid [h,x] = \alpha(h)x$ for all $h \in \mathfrak{h}\}$. We have the root space decomposition $\mathfrak{g} = \oplus\sum_{\alpha \in Q} \mathfrak{g}_\alpha$. A <u>root</u> is an element of $\Delta = \{\alpha \in Q \mid \alpha \neq 0$ and $\mathfrak{g}_\alpha \neq (0)\}$, and a <u>positive root</u> is an element of $\Delta^+ =$

$\Delta \cap Q^+$. Let $\leq$ be the partial order on $\mathfrak{h}^*$ given by $\mu \leq \lambda$ if $\lambda - \mu$ is in Q^+. Then α is in Δ^+ if and only if $\alpha > 0$ and $\mathfrak{g}_\alpha \neq (0)$. Also $\mathfrak{h} = \mathfrak{g}_0$, $\mathfrak{g}_{\alpha_i} = \mathbb{C}e_i$, $\mathfrak{g}_{-\alpha_i} = \mathbb{C}f_i$, and $\mathfrak{n}^+ = \oplus\sum_{\alpha \in \Delta^+} \mathfrak{g}_\alpha$.

For each $i = 1,\ldots,\ell$, let r_i denote the <u>fundamental reflection</u> on $\mathfrak{h}^*$ given by $r_i(\lambda) = \lambda - \lambda(h_i)\alpha_i$ for $\lambda \in \mathfrak{h}^*$. The subgroup of $GL(\mathfrak{h}^*)$ generated by the fundamental reflections is the <u>Weyl group</u> $\mathfrak{W}$ of $\mathfrak{g}$. To avoid confusion later on, we will use $\mathbb{1}$ to denote the identity element of $\mathfrak{W}$. We have $r_i^2 = \mathbb{1}$ and $r_i(\alpha_i) = -\alpha_i$. We also have that $\mathfrak{W}$ is a Coxeter group, using $\{r_1,\ldots,r_\ell\}$ as the generators. Therefore we have a length function, $\ell(w)$, defined for $w \in \mathfrak{W}$. If $\mathbf{w} = (r_{j_1},\ldots,r_{j_t})$ is a sequence of fundamental reflections such that $r_{j_1}\cdots r_{j_t} = w$ and $t = \ell(w)$, then $\mathbf{w}$ will be called a <u>reduced expression</u> for w. Since it is a Coxeter group, $\mathfrak{W}$ satisfies the <u>exchange condition</u>: If $\mathbf{w} = (r_{j_1},\ldots,r_{j_t})$ is a reduced expression for $w \in \mathfrak{W}$, and $\ell(r_i w) < \ell(w)$, then there is some s, $1 \leq s \leq t$, such that $r_{j_1}\cdots r_{j_s} = r_i r_{j_1}\cdots r_{j_{s-1}}$.

Let (V,π) be a representation of $\mathfrak{g}$. We will frequently want to view V as a $\mathfrak{g}$-module, where the action is $x\cdot v = \pi(x)(v)$ for $x \in \mathfrak{g}$ and $v \in V$. In the case of the adjoint representation $(\mathfrak{g},\mathrm{ad})$, where $\mathrm{ad}(x)(y) = [x,y]$, we will write ad_x for $\mathrm{ad}(x)$ to reduce the number of parentheses.

Given any representation (V,π) and any $\mu \in \mathfrak{h}^*$, let V_μ be the <u>weight space</u> $\{v \in V \mid h\cdot v = \mu(h)v$ for all $h \in \mathfrak{h}\}$. An element of a weight space is called a <u>weight vector</u>, and an element $\mu \in \mathfrak{h}^*$ with $V_\mu \neq (0)$ is called a <u>weight</u> of V. The representation (V,π) is $\mathfrak{h}$-<u>diagonalizable</u> if $V = \oplus\sum_{\mu \in \mathfrak{h}^*} V_\mu$. An $\mathfrak{h}$-diagonalizable representation (V,π) is called <u>integrable</u> if $\pi(e_i)$ and $\pi(f_i)$ are locally nilpotent on V for $i = 1,\ldots,\ell$ (where A acts locally nilpotently on V provided that for any $v \in V$ there is a positive integer N such that $A^N(v) = 0$). All representations considered in this paper will be integrable. In particular, the adjoint representation is integrable, and its weight spaces are just the root spaces.

Let (V,π) be an integrable representation. If there exists a nonzero weight vector v^+ such that $\mathfrak{n}^+\cdot v^+ = (0)$ and v^+ generates V as a $\mathfrak{g}$-module, then (V,π) is a <u>highest weight representation</u>. The vector v^+ is the <u>highest weight vector</u>, and if v^+ is in the weight space V_λ, then λ is called the <u>highest weight</u>

for V. Let $P^+ = \{\lambda \in \mathfrak{h}^* \mid \lambda(h_i) \in \mathbb{Z}^{\geq 0}$ for $i = 1,\ldots,\ell\}$, the lattice of dominant weights. If (V,π) is an integrable highest weight representation of highest weight λ, then λ will be in P^+.

If A is any locally nilpotent operator on V, we can define the exponential of A by

$$\exp A = I_V + \frac{1}{1!}A + \frac{1}{2!}A^2 + \cdots.$$

In particular, since $\pi(e_i)$ and $\pi(-f_i)$ are locally nilpotent on V, we can define an invertible linear transformation r_i^π of V by

$$r_i^\pi = (\exp \pi(e_i))(\exp \pi(-f_i))(\exp \pi(e_i)).$$

Finally, we let $\mathfrak{U}(\mathfrak{g})$ denote the universal enveloping algebra of $\mathfrak{g}$. We use the fact that any representation of $\mathfrak{g}$ can be extended to a unique representation of $\mathfrak{U}(\mathfrak{g})$.

2. General Results

Let (V,π) be an integrable representation of $\mathfrak{g}$. There are a few general results which we need for our study of Schubert submodules, mostly involving properties of the maps r_i^π.

Lemma 2.1. (a) For any $\mu \in \mathfrak{h}^*$, $r_i^\pi(V_\mu) = V_{r_i(\mu)}$.

(b) If β is a root and $x \in \mathfrak{g}_\beta$ is nonzero, then there is a nonzero element y of $\mathfrak{g}_{r_i(\beta)}$ such that $\pi(x)r_i^\pi = r_i^\pi\pi(y)$.

Proof. (a) This is [K], Lemma 3.8 a).
(b) If we consider the adjoint representation, then part (a) says that for any root γ, $r_i^{ad}(\mathfrak{g}_\gamma) = \mathfrak{g}_{r_i(\gamma)}$. In particular, $r_i^{ad}(\mathfrak{g}_{r_i(\beta)}) = \mathfrak{g}_\beta$. Thus, there is an element $y \in \mathfrak{g}_{r_i(\beta)}$ such that $r_i^{ad}(y) = x$, and $x \neq 0$ implies $y \neq 0$. Formula (3.8.1) of [K] says that for any locally nilpotent operator A on V and any operator B on V such that $(ad_A)^N B = 0$ for some N, the identity

$$(\exp A)B(\exp A)^{-1} = (\exp ad_A)(B)$$

holds. We may apply this formula three times to obtain

$$r_i^\pi\pi(y)(r_i^\pi)^{-1} = \exp\pi(e_i)\exp\pi(-f_i)\exp\pi(e_i)\pi(y)(\exp\pi(e_i))^{-1}(\exp\pi(-f_i))^{-1}(\exp\pi(e_i))^{-1}$$

$$= \pi\big(\exp \mathrm{ad}_{e_i} \exp \mathrm{ad}_{-f_i} \exp \mathrm{ad}_{e_i}(y)\big) = \pi(r_i^{\mathrm{ad}}(y)) = \pi(x).$$

Thus $r_i^\pi \pi(y)(r_i^\pi)^{-1} = \pi(x)$, or $r_i^\pi \pi(y) = \pi(x) r_i^\pi$, as desired. □

Corollary 2.2. Let $j_1,\ldots,j_t$ be a sequence of elements from $\{1,\ldots,\ell\}$.

(a) For any $\mu \in \mathfrak{h}^*$, $r_{j_1}^\pi \cdots r_{j_t}^\pi(V_\mu) = V_{r_{j_1}\cdots r_{j_t}(\mu)}$.

(b) If β is a root and $x \in \mathfrak{g}_\beta$ is nonzero, then there is a nonzero element y of $\mathfrak{g}_{r_{j_t}\cdots r_{j_1}(\beta)}$ such that

$$\pi(x) r_{j_1}^\pi \cdots r_{j_t}^\pi = r_{j_1}^\pi \cdots r_{j_t}^\pi \pi(y).$$

Proof. Both parts follow from the above lemma by induction on t. □

Lemma 2.3. Let W be a $\mathfrak{g}_{(i)}$-submodule of V. If W contains a nonzero element of weight μ, then it also contains a nonzero element of weight $r_i(\mu)$.

Proof. Since W is a $\mathfrak{g}_{(i)}$-module, it is an invariant subspace for $\pi(e_i)$ and $\pi(f_i)$. Thus $r_i^\pi(W) \subseteq W$. Let $v \in W \cap V_\mu$ be nonzero. Then $r_i^\pi(v)$ is nonzero and is contained in $W \cap V_{r_i(\mu)}$. □

3. Schubert Submodules

Let (V,π) be an integrable highest weight representation, with highest weight $\lambda \in P^+$. For any $w \in \mathcal{W}$, [K], Proposition 3.7 a), tells us that the weight space $V_{w(\lambda)}$ is one-dimensional. Therefore, if we choose a nonzero element $v_w \in V_{w(\lambda)}$, then any element of $V_{w(\lambda)}$ will be a scalar multiple of v_w.

Definition 3.1. For any $w \in \mathcal{W}$, we define the <u>Schubert</u> <u>submodule</u> V_w to be the $\mathfrak{b}$-submodule of V generated by v_w.

Note that V_w is independent of our choice of nonzero element v_w in V_w. In particular, if v^+ is a highest weight vector for V and $\mathbf{w} = (r_{j_1},\ldots,r_{j_t})$ is a reduced expression for w, then $v_{\mathbf{w}} = r_{j_1}^\pi \cdots r_{j_t}^\pi(v^+)$ generates V_w, since by Corollary 2.2 it is a nonzero element of $V_{w(\lambda)}$.

Our first result on Schubert submodules shows that they are finite-dimensional, even though V may be infinite-dimensional.

Lemma 3.2. For any $w \in \mathfrak{W}$, the Schubert submodule V_w is finite-dimensional.

Proof. Since it is a $\mathfrak{b}$-module, V_w is also an $\mathfrak{h}$-module, so by [K], Proposition 1.5, $V_w = \oplus\sum_{\mu\in\mathfrak{h}^*}(V_w \cap V_\mu)$. As remarked in [K], p.115, each weight space V_μ is finite-dimensional, so it suffices to show that $V_w \cap V_\mu \neq (0)$ for only a finite number of $\mu \in \mathfrak{h}^*$. Now, $V_\mu \neq (0)$ only when $\mu \le \lambda$; on the other hand, since V_w is the $\mathfrak{b}$-module generated by v_w, and v_w is of weight $w(\lambda)$, $V_w \cap V_\mu \neq (0)$ only for $\mu \ge w(\lambda)$. Thus $V_w \cap V_\mu \neq (0)$ implies $w(\lambda) \le \mu \le \lambda$, which is satisfied by only a finite number of elements μ of $\mathfrak{h}^*$. □

A more important property of Schubert submodules arises when we consider what happens when multiplying the element w of $\mathfrak{W}$ by r_i increases its length.

Proposition 3.3. Suppose that $i \in \{1,\dots,\ell\}$ and $w \in \mathfrak{W}$ have the property that $\ell(r_i w) > \ell(w)$. Then the Schubert submodule $V_{r_i w}$ is a $\mathfrak{g}_{(i)}$-module.

Proof. Let $(r_{j_1},\dots,r_{j_t})$ be a reduced expression for w. Then $\mathbf{w} = (r_i, r_{j_1},\dots,r_{j_t})$ is a reduced expression for $r_i w$, and thus $v_{\mathbf{w}} = r_i^\pi r_{j_1}^\pi \cdots r_{j_t}^\pi(v^+)$ generates $V_{r_i w}$. We claim first that $f_i \cdot v_{\mathbf{w}} = 0$.

By Corollary 2.2, there is an element x of $\mathfrak{g}_{r_{j_t}\cdots r_{j_1} r_i(-\alpha_i)}$ such that

$$\begin{aligned} f_i \cdot v_{\mathbf{w}} &= \left(\pi(f_i) r_i^\pi r_{j_1}^\pi \cdots r_{j_t}^\pi\right)(v^+) \\ &= \left(r_i^\pi r_{j_1}^\pi \cdots r_{j_t}^\pi \pi(x)\right)(v^+) \\ &= r_i^\pi r_{j_1}^\pi \cdots r_{j_t}^\pi(x \cdot v^+). \end{aligned}$$

Now $(r_i, r_{j_1},\dots,r_{j_t})$ reduced implies that $(r_{j_t},\dots,r_{j_1},r_i)$ is also reduced, and thus by [K], Lemma 3.11 b), $r_{j_t}\cdots r_{j_1} r_i(-\alpha_i) > 0$. Hence x is in $\mathfrak{n}^+$, so that $x \cdot v^+ = 0$, and $f_i \cdot v_{\mathbf{w}} = 0$, as claimed.

Next, note that because $\mathfrak{b} = \mathfrak{n}^+ \oplus \mathfrak{h}$, $\mathfrak{U}(\mathfrak{b}) = \mathfrak{U}(\mathfrak{n}^+) \otimes \mathfrak{U}(\mathfrak{h})$. Since $V_{r_i w} = \mathfrak{U}(\mathfrak{b}) \cdot v_{\mathbf{w}}$ and $v_{\mathbf{w}}$ is a weight vector, we have $V_{r_i w} = \mathfrak{U}(\mathfrak{n}^+) \cdot v_{\mathbf{w}}$. But $\mathfrak{U}(\mathfrak{n}^+)$ is generated (as an associative algebra) by $e_1,\dots,e_\ell$, so this shows that $V_{r_i w}$ is spanned by the set

$$(3.1) \qquad \{e_{k_1} \cdots e_{k_m} \cdot v_{\mathbf{w}} \mid m \ge 0,\ k_1,\dots,k_m \in \{1,\dots,\ell\}\}.$$

We can now show that V_{r_iw} is closed under the action of f_i, which will be sufficient to demonstrate that it is a $\mathfrak{g}_{(i)}$-module. From the spanning set (3.1), it is enough to prove that $f_i\cdot(e_{k_1}\cdots e_{k_m}\cdot v_w)$ is in V_{r_iw} for any $m \geq 0$ and any $k_1,\ldots,k_m \in \{1,\ldots,\ell\}$, which we do by induction on m.

For $m = 0$, we have $f_i\cdot v_w = 0$, which is certainly in V_{r_iw}. Now suppose the result holds for some m and consider $f_i\cdot(e_{k_1}e_{k_2}\cdots e_{k_{m+1}}\cdot v_w)$. This equals

$$e_{k_1}f_ie_{k_2}\cdots e_{k_{m+1}}\cdot v_w + [f_i,e_{k_1}]e_{k_2}\cdots e_{k_{m+1}}\cdot v_w.$$

By induction $f_ie_{k_2}\cdots e_{k_{m+1}}\cdot v_w$ is in V_{r_iw}, which is a $\mathfrak{b}$-module, so $e_{k_1}f_ie_{k_2}\cdots e_{k_{m+1}}\cdot v_w$ is in V_{r_iw}. For the second term, note that $[f_i,e_{k_1}]$ is either $-h_i$ or 0, both of which are elements of $\mathfrak{h}$, and thus $[f_i,e_{k_1}]e_{k_2}\cdots e_{k_{m+1}}\cdot v_w$ is in V_{r_iw}. Hence $f_i\cdot(e_{k_1}e_{k_2}\cdots e_{k_{m+1}}\cdot v_w)$ is in V_{r_iw} also.

Therefore we have that V_{r_iw} is closed under the action of f_i. Since e_i and h_i are in $\mathfrak{b}$, V_{r_iw} is closed under the action of these elements also, so V_{r_iw} is a $\mathfrak{g}_{(i)}$-module. □

Corollary 3.4. *If $w \in \mathcal{W}$ is such that $\ell(r_iw) > \ell(w)$, then $V_w \subsetneq V_{r_iw}$.*

Proof. It suffices to show that v_w is in V_{r_iw}, and since v_w is in the weight space $V_{w(\lambda)}$, which is one-dimensional, it is enough to show that V_{r_iw} contains a nonzero element of weight $w(\lambda)$. But this follows immediately from the proposition just proved and Lemma 2.3.□

Lemma 3.5. ([Hu], Lemma 21.2) For $k \geq 0$ and $1 \leq i,j \leq \ell$, the following identities hold in $\mathfrak{U}(\mathfrak{g})$:

(i) $[e_j,f_i^k] = 0$ for $i \neq j$;

(ii) $[e_i,f_i^k] = k(1-k)f_i^{k-1} + kf_i^{k-1}h_i$;

(iii) $[h,f_i^k] = -k\alpha_i(h)f_i^k$ for any $h \in \mathfrak{h}$.

Proposition 3.6. *Let w be in $\mathcal{W}$ and suppose $i \in \{1,\ldots,\ell\}$ is such that $\ell(r_iw) > \ell(w)$. Then V_{r_iw} is the $\mathfrak{g}_{(i)}$-submodule of V generated by V_w.*

Proof. Let W be the $\mathfrak{g}_{(i)}$-submodule of V generated by V_w. Then by

Corollary 3.4 and Proposition 3.3, $W \subseteq V_{r_i w}$. On the other hand, $W = \mathfrak{U}(\mathfrak{g}_{(i)})\cdot V_w$, and the Poincaré-Birkhoff-Witt Theorem, coupled with the fact that V_w is a $\mathfrak{b}$-module, shows that the set $\{f_i^k \cdot v \mid k \geq 0 \text{ and } v \in V_w\}$ spans W. The preceding lemma shows that for any $k \geq 0$ and any $v \in V_w$, $e_j\cdot(f_i^k\cdot v)$ and $h\cdot(f_i^k\cdot v)$ are in W for $j = 1,\ldots,\ell$ and for any $h \in \mathfrak{h}$. Since $\mathfrak{b}$ is generated by $\mathfrak{h}$ and $e_1,\ldots,e_\ell$, W is a $\mathfrak{b}$-module. Thus, to show $W = V_{r_i w}$, we need only find a nonzero element of weight $r_i w(\lambda)$ in W. Lemma 2.3 reduces the problem to that of finding a nonzero element of weight $w(\lambda)$ in W. This is easily solved, though, for v_w is in V_w and $V_w \subseteq W$. □

Corollary 3.7. For any $w \in W$ and any $i \in \{1,\ldots,\ell\}$, the $\mathfrak{g}_{(i)}$-submodule of V generated by V_w is a Schubert submodule.

Proof. If $\ell(r_i w) > \ell(w)$, then the $\mathfrak{g}_{(i)}$-submodule of V generated by V_w is $V_{r_i w}$ by the proposition just proved. If $\ell(r_i w) < \ell(w)$, then the exchange condition for elements of $\mathfrak{W}$ shows that w has a reduced expression of the form $(r_i, r_{j_1}, \ldots, r_{j_t})$. By Proposition 3.3, V_w itself is a $\mathfrak{g}_{(i)}$-module, so the $\mathfrak{g}_{(i)}$-submodule of V generated by V_w is just V_w. □

Let us now relate a restricted Bruhat ordering on $\mathfrak{W}$ to the inclusion relation on Schubert submodules.

Definition 3.8. For $w, w' \in \mathfrak{W}$, we write $w \to w'$ if $w' = r_i w$ for some $i \in \{1,\ldots,\ell\}$ and $\ell(w') > \ell(w)$. We write $w \leq w'$ if $w = w'$ or if there is a sequence of elements $w = w_1 \to w_2 \to \cdots \to w_n = w'$ in $\mathfrak{W}$.

This is a restriction of the usual Bruhat order on $\mathfrak{W}$, in that we only allow multiplication on the left by fundamental reflections, whereas the usual ordering multiplication by conjugates of fundamental reflections.

Lemma 3.9. If w and w' are in $\mathfrak{W}$ and $w \leq w'$, then $V_w \subset V_{w'}$.

Proof. This follows immediately from the definition of the ordering and Corollary 3.4. □

As we shall see later, if V is finite-dimensional, then V itself is a Schubert submodule. If V is infinite-dimensional, V cannot be a Schubert submodule (since Schubert submodules are finite-dimensional). However, we can still obtain V as a union of Schubert submodules in this case. To prove this, we first need the notion of a directed set.

Definition 3.10. Let D be a partially ordered set. We call D a directed set if for any $\delta, \delta' \in D$ there is some $\delta^* \in D$ with $\delta \leq \delta^*$ and $\delta' \leq \delta^*$.

Lemma 3.11. Let $\mathcal{S}$ be the collection of all Schubert submodules of V, partially ordered by inclusion. Then $\mathcal{S}$ is a directed set.

Proof. Let V_w and $V_{w'}$ be two Schubert submodules of V. Let $(r_{j_t},\ldots,r_{j_1})$ be a reduced expression for w', and set $W_0 = V_w$, W_k = the $\mathfrak{g}(j_k)$-submodule of V generated by W_{k-1} for $k = 1,\ldots,t$. Clearly $W_0 \subseteq W_1 \subseteq \ldots \subseteq W_t$, and by Corollary 3.7, each W_k is a Schubert submodule. In addition, set $\sigma_0 = \mathbf{1}$, $\sigma_k = r_{j_k}\cdots r_{j_1}$ for $k = 1,\ldots,t$, and note that $\sigma_k = r_{j_k}\sigma_{k-1}$. We claim that $V_{\sigma_k} \subseteq W_k$ for $k = 0,1,\ldots,t$.

Induct on k. For $k = 0$, we must show that $V_{\mathbf{1}} \subseteq W_0 = V_w$. But $\mathbf{1} \leq w$ in our ordering on $\mathcal{W}$, so by Lemma 3.9 $V_{\mathbf{1}} \subseteq V_w$, as desired. Now suppose that $V_{\sigma_{k-1}} \subseteq W_{k-1}$. Since $\ell(r_{j_k}\sigma_{k-1}) > \ell(\sigma_{k-1})$, V_{σ_k} is the $\mathfrak{g}(j_k)$-submodule of V generated by $V_{\sigma_{k-1}}$. Thus V_{σ_k} is contained in the $\mathfrak{g}(j_k)$-submodule of V generated by W_{k-1}, which is W_k. That is, $V_{\sigma_k} \subseteq W_k$.

Now we have that W_t is a Schubert submodule, $V_w = W_0 \subseteq W_t$ and $V_{w'} = V_{\sigma_t} \subseteq W_t$. This shows that $\mathcal{S}$ is a directed set. □

To help illustrate the significance of Lemma 3.11, we point out that $\mathcal{W}$ need not be a directed set under our partial ordering. This can be seen fairly quickly when $\mathfrak{g}$ is the algebra $A_1^{(1)}$, which has Cartan matrix $A = \begin{bmatrix} 2 & -2 \\ -2 & 2 \end{bmatrix}$. The Weyl group of $\mathfrak{g}$ is an infinite dihedral group. Thus, if we set

$$w_{2k,1} = (r_2r_1)^k, \qquad w_{2k+1,1} = r_1(r_2r_1)^k,$$

$$w_{2k,2} = (r_1r_2)^k, \qquad w_{2k+1,2} = r_2(r_1r_2)^k,$$

then the discussion in [B], page 11, shows that these elements $(w_{1,1}, w_{2,1}, w_{3,1}, \ldots, w_{1,2}, w_{2,2}, w_{3,2}, \ldots)$ are all distinct, and that each of them has a unique reduced expression, given by $(\ldots, r_1, r_2, r_1)$ for $w_{n,1}$ and $(\ldots, r_2, r_1, r_2)$ for $w_{n,2}$.

An easy consequence of Definition 3.8 is that $w \leq w'$ if and only if any reduced expression $(r_{j_1}, \ldots, r_{j_t})$ for w can be extended on the left to a reduced expression $(r_{i_1}, \ldots, r_{i_m}, r_{j_1}, \ldots, r_{j_t})$ for w'. From this observation it follows that the set of non-identity elements of $\mathfrak{W}$ splits into two linearly ordered subsets, $w_{1,1} < w_{2,1} < w_{3,1} < \ldots$ and $w_{1,2} < w_{2,2} < w_{3,2} < \ldots$, and that $w_{n,1}$ and $w_{n',2}$ are not related for any n and n' in $\mathbb{Z}^+$. In particular, $\mathfrak{W}$ is not a directed set, because there is no element w with $w_{1,1} \leq w$ and $w_{1,2} \leq w$.

Proposition 3.12. Let $\mathcal{S}$ be the collection of all Schubert submodules of V. Then $\cup\mathcal{S} = V$.

Proof. Let $W = \cup\mathcal{S}$. Note that $\mathbf{1}(\lambda) = \lambda$, so we may take $v_{\mathbf{1}}$ to be v^+, a highest weight vector for V. From this it is clear that the Schubert submodule $V_{\mathbf{1}} = \mathbb{C}v^+$; in particular, W contains v^+. Thus, to show that $W = V$, it suffices to show that W is a $\mathfrak{g}$-module.

Because each Schubert submodule is a $\mathfrak{b}$-module and $\mathcal{S}$ forms a directed set under inclusion, it is easy to demonstrate that $W = \cup\mathcal{S}$ is a $\mathfrak{b}$-module. To show that W is a $\mathfrak{g}$-module, then, it is enough to prove that W is closed under the actions of $f_1, \ldots, f_\ell$, since these elements, together with $\mathfrak{b}$, generate $\mathfrak{g}$.

Fix $i \in \{1, \ldots, \ell\}$, and let v be in W, so that v is in V_w for some $w \in \mathfrak{W}$. Then $f_i \cdot v$ is in the $\mathfrak{g}_{(i)}$-submodule of V generated by V_w, which, according to Corollary 3.7, is another Schubert submodule. Thus $f_i \cdot v$ is in W, so W is closed under the action of f_i. This holds for all $i \in \{1, \ldots, \ell\}$, which shows that W is a $\mathfrak{g}$-module, and therefore that $W = V$. □

Corollary 3.13. If V is finite-dimensional, then V is a Schubert submodule.

Proof. Let $v_1,\dots,v_n$ be a basis for V. Since $V = \cup\mathcal{S}$, there are elements $w_1,\dots,w_n \in \mathcal{W}$ such that v_i is in V_{w_i} for $i = 1,\dots,n$. But $\mathcal{S}$ is a directed set, so there is a Schubert submodule V_{w^*} with $V_{w_i} \subseteq V_{w^*}$ for $i = 1,\dots,n$. Then v_i is in V_{w^*} for each i, which implies that $V = V_{w^*}$. □

If $\mathfrak{g}$ is a finite-dimensional semisimple Lie algebra, then $\mathcal{W}$ has a unique element w_0 of maximum length (see [B], Chapter VI, §1, n° 6). It follows from [H], Corollary 1.4, that $w \leq w_0$ for all $w \in \mathcal{W}$. Since any integrable highest weight module V for $\mathfrak{g}$ is finite-dimensional in this case, the above corollary, together with Lemma 3.9, implies that $V = V_{w_0}$. For a slightly different proof of this fact, see Chapter 2 of [H].

Returning to the situation that $\mathfrak{g}$ is an arbitrary Kac-Moody algebra, we note that it is not always necessary to take a union of all Schubert submodules in order to obtain V. Our final result describes one situation in which it is possible to obtain V as a union of some subcollection of the set of all Schubert submodules.

Definition 3.14. Let i be in $\{1,\dots,\ell\}$. We will call a nonempty subset $\mathcal{J} \subseteq \mathcal{W}$ r_i-cofinal if for any $w \in \mathcal{J}$ there is some $w' \in \mathcal{J}$ such that $w \leq w'$ and w' has a reduced expression of the form $(r_i, r_{j_t},\dots,r_{j_1})$.

Proposition 3.15. Let $\mathcal{J}$ be a subset of $\mathcal{W}$ which is a directed set and is r_i-cofinal for all $i = 1,\dots,\ell$. Then $\bigcup_{w\in\mathcal{J}} V_w = V$.

Proof. Let $W = \bigcup_{w\in\mathcal{J}} V_w$. As in the proof of Proposition 3.12, the fact that $\mathcal{J}$ is a directed set insures that W is a $\mathfrak{b}$-module. Moreover, if we pick any $w \in \mathcal{J}$, we have $\mathbf{1} \leq w$, so $V_{\mathbf{1}} \subseteq V_w \subseteq W$, and therefore W contains v^+, a highest weight vector for V. Thus, as above, it suffices to show that W is closed under the actions of $f_1,\dots,f_\ell$.

Fix $i \in \{1,\dots,\ell\}$ and let v be in W, so that v is in V_w for some $w \in \mathcal{J}$. Let w' be in $\mathcal{J}$ with $w \leq w'$ and w' having a reduced expression of the form $(r_i, r_{j_t},\dots,r_{j_1})$. Then $V_w \subseteq V_{w'}$ and $V_{w'}$ is a $\mathfrak{g}_{(i)}$-module, so $f_i\cdot v$ is an element of $V_{w'}$, and hence also of W. This shows that W is closed under the action of

f_i, and therefore that $W = V$. □

We can give a quick illustration of how Proposition 3.15 might be applied using $A_1^{(1)}$ again. Using the notation introduced after Lemma 3.11, let $\mathcal{J} = \{w_{1,1}, w_{2,1}, w_{3,1}, \ldots\}$. As we pointed out at that time, the elements of $\mathcal{J}$ are linearly ordered, so certainly $\mathcal{J}$ is a directed set. Furthermore, the reduced expression for $w_{n,1}$ begins with r_1 if n is odd and r_2 if n is even, so $\mathcal{J}$ is r_1- and r_2-cofinal. Thus, Proposition 3.15 tells us that for any integrable highest weight module V of $A_1^{(1)}$,

$$V = \bigcup_{w \in \mathcal{J}} V_w = \bigcup_{n=1}^{\infty} V_{w_{n,1}}.$$

References

[B] N. Bourbaki, *Groupes et algèbres de Lie, Chapitres IV, V et VI*, Hermann, Paris (1968).

[H] M. Hall, *Verma Bases of Modules for Simple Lie Algebras*, PhD thesis, University of Wisconsin-Madison (1987).

[Hu] J. E. Humphreys, *Introduction to Lie Algebras and Representation Theory*, Springer-Verlag, New York (1972).

[J] A. Joseph, On the Demazure character formula, Annales Scientifiques de l'École Normale Superieure, 4ᵉ serie, **18**(1985), 389-419.

[K] V. Kac, *Infinite-dimensional Lie Algebras*, 2nd edition, Cambridge University Press, Cambridge (1985).

[Ku] S. Kumar, Demazure character formula in arbitrary Kac-Moody setting, Inventiones mathematicae, **89**(1987), 395-423.

[LMNP] S.-P. Li, R. V. Moody, M. Nicolescu, J. Patera, Verma bases for representations of classical simple Lie algebras, Journal of Mathematical Physics, **27**(1986), 668-677.

1980 AMS subject classification (1985 revision): 17B67

Division of Mathematics and Computer Science
Northeast Missouri State University
Kirksville, MO 63501-0828

This paper is in final form and no version of it will appear elsewhere.

KAC–MOODY MODULES AND GENERALIZED CLIFFORD ALGEBRAS

George B. Seligman*

Abstract. It is shown that all the irreducible integrable highest–weight modules for a Kac–Moody algebra of type $C_\ell^{(1)}$, and a large class of those for type $D_\ell^{(1)}$, can be constructed by essentially the same formalism as that employed for certain fundamental modules by Frenkel, Kac–Peterson and Feingold–Frenkel. The idea is to replace the Clifford algebra (or Weyl algebra) by higher analogues of the even Clifford algebra, and the spin, half–spin, or oscillator module by a suitable simple module for this (locally finite) algebra. A submodule of the resulting completely reducible module for the Kac–Moody algebra then affords the desired representation. Properties of the larger module, and analogy with cases previously investigated, suggested the conjecture that our submodule is (often) not proper. In general, this is not the case, as is seen by comparison of characters for certain "basic" examples. The determination of the character of our module rests on finding explicit bases.

§1. The Lie Algebras and a Construction of Modules.

Let V be a (non–zero) finite–dimensional complex vector space with a nondegenerate bilinear form (u,v), which shall be either symmetric $(\varepsilon = 1)$ or antisymmetric $(\varepsilon = -1)$. We fix a basis for V on which the inner product is canonical:

If $\dim V = 2\ell$: $e_{-\ell},\dots,e_{-1};e_1,\dots,e_\ell$, with

$$(e_i,e_j) = 0 \text{ if } i \neq j; \quad (e_{-j},e_j) = 1 = \varepsilon(e_j,e_{-j}),\ 1 \leq j \leq \ell .$$

If $\dim V = 2\ell + 1$: $e_{-\ell},\dots,e_{-1},e_0,e_1,\dots,e_\ell$, with $(e_i,e_j) = \delta_{i,-j}$.

Let $\mathfrak{g}$ be the Lie algebra of linear transformations of V leaving the form infinitesimally invariant, viewed as acting on the right: $T \in \mathfrak{g}$ means $(uT,v) + (u,vT) = 0$ for all $u,v \in V$.

Let t be an indeterminate over $\mathbb{C}$, and form the $\mathbb{C}$–vector space

---

*Research at Yale University and I.H.E.S., Bures–sur–Yvette, supported in part by National Science Foundation Grants Nos. MCS82–01333 and DMS–8512904.

$$B = (g \otimes_{\mathbb{C}} \mathbb{C}[t,t^{-1}]) \oplus \mathbb{C}c \oplus \sum_{n=-\infty}^{\infty} \oplus \mathbb{C}d(n) ,$$

where each of $\mathbb{C}c$, $\mathbb{C}d(n)$ $(n \in \mathbb{Z})$ is a one-dimensional space with the displayed element as basis.

We define an antisymmetric product on B, using its decomposition above and the decomposition of $g \otimes \mathbb{C}[t,t^{-1}]$ into the direct sum of the subspaces $g \otimes t^i$, $i \in \mathbb{Z}$:

i) $[X \otimes t^i, Y \otimes t^j] = [XY] \otimes t^{i+j} + i\delta_{i,-j}\mathrm{Tr}(XY)c$, for $X,Y \in g$, where the trace is that of endomorphisms of V;

ii) $[c,B] = 0 = [B,c]$;

iii) $[d(n), X \otimes t^i] = iX \otimes t^{n+i} = -[X \otimes t^i, d(n)]$;

iv) $[d(n),d(r)] = (r-n)d(n+r) - \delta_{n,-r}\frac{\gamma}{6}(n^3-n)c$.

Here γ is a non-zero parameter to be specified later. (If $\dim V = 2\ell$, we shall take $\gamma = \ell$.) It is a perfectly routine matter to verify the Jacobi identity for the product, so that B is a Lie algebra over $\mathbb{C}$.

The subspace $D = \mathbb{C}c + \sum_{n\in\mathbb{Z}} \mathbb{C}d(n)$ is a subalgebra, the Virasoro algebra, with $\mathbb{C}c$ as center, $D/\mathbb{C}c \cong \mathrm{Der}_{\mathbb{C}}\mathbb{C}[t,t^{-1}]$, and iii) expresses the fact that the adjoint action of D on $g \otimes \mathbb{C}[t,t^{-1}]$ is the canonical action of $\mathrm{Der}\,\mathbb{C}[t,t^{-1}]$ on the second factor. The subspace

$$K = g \otimes [t,t^{-1}] \oplus \mathbb{C}c \oplus \mathbb{C}d(0)$$

is also a subalgebra. When g is simple (thus except when $\varepsilon = 1$ and $\dim V = 1, 2$ or 4), we associate with g the extended Dynkin diagram as follows:

$\varepsilon = 1$, $\dim V = 2\ell > 6$;

; $\dim V = 6$:

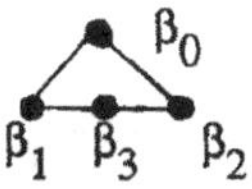

$\varepsilon = 1$, $\dim V = 2\ell+1 > 5$;

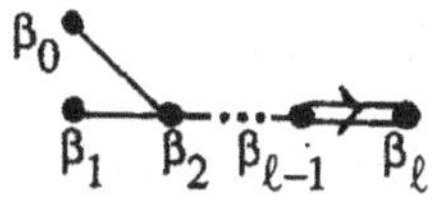

; $\dim V = 5$:

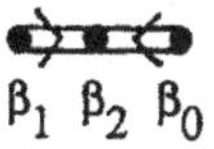

$\dim V = 3$:

$\varepsilon = -1$, $\dim V = 2\ell > 2$:

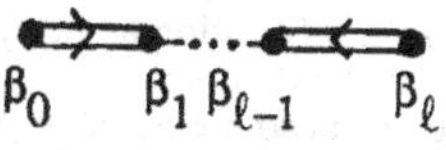

; $\dim V = 2$:

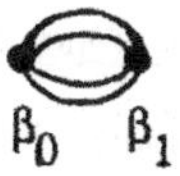

Let $A = (a_{ij})$, $0 \le i,j \le \ell$ be the associated (generalized) Cartan matrix, which is symmetrizable in the sense of [4], §2.1. Then $K \equiv \mathfrak{g}(A)$ in the sense of [4], §1.3, and the derived Lie algebra

$$\mathfrak{g}'(A) \equiv K' = \mathfrak{g} \otimes \mathbb{C}[t,t^{-1}] \oplus \mathbb{C}c\,.$$

Let $\Lambda^{-2}(V)$ denote the symmetric product $S^2(V)$, with $\Lambda^2(V)$ as usual. Then the bilinear pairing sending u,v to $S_{u,v} \in \mathfrak{g}$: $wS_{u,v} = (w,u)v - \varepsilon(w,v)u$ induces a linear isomorphism of $\Lambda^{2\varepsilon}(V)$ onto $\mathfrak{g}$.

Now let $V_\infty = V \otimes_{\mathbb{C}} \mathbb{C}[t,t^{-1}] = \sum_{i \in \mathbb{Z}} \oplus (V \otimes t^i)$, with inner product $(u \otimes t^i, v \otimes t^j) = \delta_{i,-j}(u,v)$, extended uniquely by biadditivity. Then $V_0 = V \otimes t^0$ is isomorphic to V as space with bilinear form, and V_∞ is a split infinite-dimensional space with splitting basis

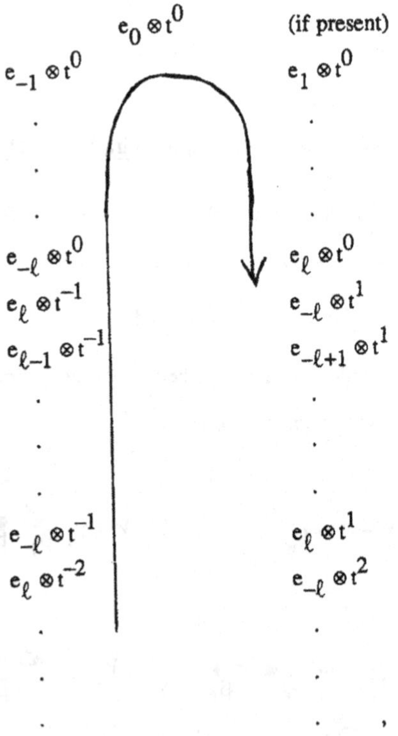

where the curved arrow indicates the direction of progression from $-\infty$ to $+\infty$. Taking $n_0 = \ell$, $n_r = \ell + r(\dim V)$ for $r = 1,2,\ldots$, we have a filtration as in [6]: V_r = span of those basis vectors in the canonical basis above, with index of absolute value at most n_r; that is, $V_r =$

$\sum_{|j|\leq n_r} V\otimes t^j$, and all the considerations of [6] apply to this situation. We fix a positive integer k, with k even if $\varepsilon=-1$, and accordingly the locally finite generalized even Clifford algebra $A_k = A_k(V_\infty)$, with canonical map $\varphi: V_\infty \otimes V_\infty \to A_k$ [6]. Our identification of V_0 with V enables us to label consistently the infinite basis, in order as given above, as $\{e_i \mid i \in \mathbb{Z}$, if $\dim V$ is odd$\}$, or as $\{e_i \mid i \in \mathbb{Z}-\{0\}$, if $\dim V$ is even$\}$, with $e_i = e_i \otimes t^0$, $-\ell \leq i \leq \ell$.

With k fixed as above, let $\Lambda = (m_1, m_2, \ldots)$ be an infinite sequence of non-negative integers, with $s(\Lambda) = k$, where $s(\Lambda) = 2(m_1+m_2+\ldots)$, if $\varepsilon=-1$;

$$\begin{array}{ll} m_1+m_2+2(m_3+m_4+\ldots), & \text{if } \varepsilon=1, \quad \dim V \text{ even;} \\ m_1+2(m_2+m_3+\ldots) & \text{if } \varepsilon=1, \quad \dim V \text{ odd.} \end{array}$$

Associated with Λ we have a right ideal in A_k generated by all $\varphi(e_i \otimes e_j)$, $i+j>0$, and by all $H_i - m_i 1$, where $H_i = \varphi(e_{i-1} \otimes e_{1-i}) - \varphi(e_i \otimes e_{-i})$ for $i>1$ and: $H_1 = -\varphi(e_1 \otimes e_{-1}) + \frac{k}{2}1$, if $\varepsilon=-1$; $H_1 = -2\varphi(e_1 \otimes e_{-1}) + 1$, if $\varepsilon=1$, $\dim V$ odd; $H_1 = -\varphi(e_1 \otimes e_{-1}) - \varphi(e_2 \otimes e_{-2}) + 1$ if $\varepsilon=1$, $\dim V$ even. This ideal b_Λ is also generated by the smaller set consisting of the $H_i - m_i 1$, and by all $X_{\alpha_i} = \varphi(e_{-1} \otimes e_{i+1})$, $i>0$, together with: $X_{\alpha_i} = \varphi(e_0 \otimes e_1)$ if $\varepsilon=1$, $\dim V$ odd; $X_{\alpha_1} = \varphi(e_1 \otimes e_2)$ if $\varepsilon=1$, $\dim V$ even; $X_{\alpha_1} = \varphi(e_1 \otimes e_1)$, if $\varepsilon=-1$; and it is a maximal right ideal in A_k [6]. Thus the right quotient module $M_\Lambda = A_k/b_\Lambda$ is a simple right A_k-module; the canonical image $\bar{1}$ of $1 \in A_k$ is a generator, and b_Λ is the annihilator of $\bar{1}$.

Let $X'_{-\alpha_i}$ be obtained from X_{α_i} above by replacing the indices of the corresponding e_j's by their negatives: thus $X'_{-\alpha_1} = \varphi(e_0 \otimes e_{-1})$ if $\varepsilon=1$ and $\dim V$ is odd. Then M_Λ is linearly generated by all elements

$$\bar{1}\, X'_{-\alpha_{i_1}} \cdots X'_{-\alpha_{i_t}}, \tag{1}$$

where $(i_1,\ldots,i_t)$ runs over finite sequences of positive integers. Each of these is a common eigenvector for all $\varphi(e_j \otimes e_{-j})$, $j>0$. If c_j is the corresponding eigenvalue, we call the sequence (c_j) the A_k-weight of the element in question. Almost all c_j are equal to zero, and the c_j are all rational integers or half-integers. The space of vectors of A_k-weight (c_j) is finite-dimensional [6].

Now consider an element $m=(1)$ as above. Let $u,v \in V$ be fixed and let $n \in \mathbb{Z}$ be fixed. For $j \in \mathbb{Z}$, $j > \frac{n}{2}$, we consider the effect on m of $\varphi(u(j) \otimes v(n-j))$, where $u(j) = u \otimes t^j \in V_\infty$.

Whenever j is sufficiently large, u(j) is a linear combination of elements in our canonical basis with indices larger (in absolute value) than all those occurring in $X'_{-\alpha_{i_1}},\ldots,X'_{-\alpha_{i_t}}$, and likewise for v(n–j) (the indices here being negative, but large in absolute value). From the defining relations of A_k it follows that $\varphi(u(j) \otimes v(n-j))$ commutes with all $X'_{-\alpha_i}$ involved in our expression for m. Thus to show that $m\varphi(u(j) \otimes v(n-j)) = 0$ for all but finitely many $j > \frac{n}{2}$, it suffices to show

$$\varphi(u(j) \otimes v(n-j)) \in b_\Lambda$$

for all but finitely many $j > \frac{n}{2}$.

For $n > 0$ this follows by our comments about the generation of b_Λ. The fact that $m_j = 0$ for j sufficiently large implies that $\varphi(e_{-j-1} \otimes e_j)$ annihilates Υ for s sufficiently large, and the commutator

$$[\varphi(e_{-(s+1)} \otimes e_s), \varphi(e_{-(s+2)} \otimes e_{s+1})]$$

is a non-zero scalar multiple of $\varphi(e_{-(s+2)} \otimes e_s)$. Repeating, we see that Υ is annihilated by $\varphi(e_j \otimes e_{-(j+p)})$ for all j sufficiently large, all $p > 0$. It follows for $n < 0$ that $m\varphi(u(j) \otimes v(n-j)) = 0$ for all but finitely many $j > \frac{n}{2}$. When $n = 0$, it suffices to prove the assertion for u,v in our basis $e_{-\ell},\ldots,e_\ell$ for V, and the observations above show that the assertion holds except perhaps when $u = e_i$, $v = e_j$, $i+j = 0$. Here the element of A_k in question has the form $\varphi(e_s \otimes e_{-s})$, for $s > 0$. The effect of $\varphi(e_s \otimes e_{-s})$ on Υ is to send Υ to $c_s\Upsilon$, where (c) is the weight of Υ. From [6] (§3.b), it is known that $c_s = 0$ for s sufficiently large. Thus we have the

<u>Lemma 1</u>. Let $m \in M_\Lambda$; $u,v \in V$; $m \in \mathbb{Z}$. Then for all but finitely many $j > \frac{n}{2}$, $m\varphi(u(j) \otimes v(n-j)) = 0$.

It follows that for u,v,n as in Lemma 1 an endomorphism $f_n(u,v)$ of M_Λ is defined by setting

$$(2) \quad mf_n(u,v) = \sum_{j>\frac{n}{2}} m(\varphi(u(j) \otimes v(n-j)) - \varepsilon\varphi(v(j) \otimes u(n-j)) + \tfrac{1}{2}m(\varphi(u(\tfrac{n}{2}) \otimes v(\tfrac{n}{2})) - \varepsilon\varphi(v(\tfrac{n}{2}) \otimes u(\tfrac{n}{2}))$$

for each $m \in M_\Lambda$, the last term being present only if n is even. Clearly f_n is bilinear in u and v, and annihilates $\Lambda^{-2\varepsilon}(V)$ when viewed as a mapping $V \otimes V \to \mathrm{End}_{\mathbb{C}}(M_\Lambda)$. From

$V \otimes V = \Lambda^{-2\varepsilon}(V) \oplus \Lambda^{2\varepsilon}(V)$ and the linear isomorphism $\mathfrak{g} \cong \Lambda^{2\varepsilon}(V)$ it follows that there is a linear mapping $p_n : \mathfrak{g} \to \mathrm{End}_{\mathbb{C}}(M_\Lambda)$ with $p_n(S_{u,v}) = f_n(u,v)$. For $m \in M_\Lambda$, $X \in \mathfrak{g}$, $n \in \mathbb{Z}$, we set

$$m \cdot (X \otimes t^n) = m p_n(X)\,, \tag{3}$$

and we extend by additivity to $\mathfrak{g} \otimes \mathbb{C}[t,t^{-1}] = \sum_i \oplus (\mathfrak{g} \otimes t^i)$, obtaining a linear mapping $\mathfrak{g} \otimes \mathbb{C}[t,t^{-1}] \to \mathrm{End}_{\mathbb{C}}(M_\Lambda)$. We let c act on M_Λ by the scalar $-\frac{k}{2}$.

As above, if we fix v, $1 \le v \le \ell$, the operator $D_v(n)$:

$$m \mapsto \sum_{j>\frac{n}{2}} (j - \tfrac{n}{2}) m(\varphi(e_v(j) \otimes e_{-v}(n-j)) + \varepsilon\varphi(e_{-v}(j) \otimes e_v(n-j))) + \delta_{n,0} \frac{k}{8} m$$

is well-defined, as is the operator $D(n) = \sum_{v=1}^{\ell} D_v(n)$.

Proposition 1: When $\dim V$ is even, the linear mapping $B \to \mathrm{End}_{\mathbb{C}}(M_\Lambda)$ sending $X \otimes t^n$ to $p_n(X)$, c to $-\frac{k}{2}I$, $d(n)$ to $D(n)$ defines a structure of right B-module on M_Λ.

The proof of the proposition is a rather laborious calculation, and will be omitted. No particular subtleties are involved. (For a more subtle argument when $\varepsilon = 1$ and $k = 1$, see [3], §3.)

For the rest of the paper, $\dim V = 2\ell$ will be assumed.

§2. Weight-Space Decomposition. Integrability.

We consider the action on M_Λ of the subalgebra $K = \mathfrak{g}(A)$ of B. The canonical split toral subalgebra H of $\mathfrak{g}(A)$ has as basis the $S_{e_v,e_{-v}}$, $1 \le v \le \ell$; $d(0)$; and c. A set of root-vectors for the simple roots $\beta_0, \beta_1, \ldots, \beta_\ell$ and for their negatives is as follows:

$$x_{\beta_v} = S_{e_{-(\ell-v)}, e_{\ell-v+1}} \otimes t^0, \qquad 1 \le v < \ell\,;$$

$$x_{-\beta_v} = S_{e_{\ell-v}, e_{-(\ell-v+1)}} \otimes t^0, \qquad 1 \le v < \ell\,;$$

when $\varepsilon = -1$,

$$x_{\beta_\ell} = \tfrac{1}{2}S_{e_1,e_1} \otimes t^0, \qquad x_{\beta_0} = \tfrac{1}{2}S_{e_{-\ell},e_{-\ell}} \otimes t^1,$$

$$x_{-\beta_\ell} = \tfrac{1}{2}S_{e_{-1},e_{-1}} \otimes t^0, \qquad x_{-\beta_0} = \tfrac{1}{2}S_{e_\ell,e_\ell} \otimes t^{-1};$$

when $\varepsilon = 1$,

$$x_{\beta_\ell} = S_{e_1,e_2} \otimes t^0, \qquad x_{\beta_0} = S_{e_{-\ell},e_{-\ell+1}} \otimes t^1,$$

$$x_{-\beta_\ell} = S_{e_{-1},e_{-2}} \otimes t^0, \qquad x_{-\beta_0} = S_{e_\ell,e_{\ell-1}} \otimes t^{-1}.$$

These are so chosen that if β is a simple root,

$$[x_\beta[x_\beta x_{-\beta}]] = 2x_\beta .$$

We further have $[x_{\beta_\nu},d(0)] = 0 = [x_{-\beta_\nu},d(0)]$ unless $\nu = 0$, when $[x_{\beta_0},d(0)] = -x_{\beta_0}$, $[x_{-\beta_0},d(0)] = x_{-\beta_0}$.

We claim M_Λ is a sum of H–weight spaces. For this the element c need not be considered; we consider first the generator $\bar{1}$ of M_Λ. By our formula,

$$\bar{1}(S_{e_\nu,e_{-\nu}} \otimes t^0) = \sum_{j>0} \bar{1}(\varphi(e_\nu(j) \otimes e_{-\nu}(-j)) - \varepsilon\varphi(e_{-\nu}(j) \otimes e_\nu(-j)))$$

$$+ \tfrac{1}{2}\,\bar{1}(\varphi(e_\nu(0) \otimes e_{-\nu}(0)) - \varepsilon\varphi(e_\nu(0) \otimes e_\nu(0))) .$$

Now for $\nu > 0, j > 0$,

$$e_\nu(j) = e_{\nu+2\ell j}, \qquad e_{-\nu}(j) = e_{-\nu+2\ell j+1},$$

$$e_\nu(-j) = \varepsilon e_{\nu-2j\ell-1}, \qquad e_{-\nu}(-j) = e_{-\nu-2j\ell} .$$

Thus the above is

$$\sum_{j>1} \bar{1}(\varphi(e_{\nu+2\ell j} \otimes e_{-\nu-2\ell j}) - \varphi(e_{2\ell j+1-\nu} \otimes e_{-2\ell j-1+\nu})) + \tfrac{1}{2}\,\bar{1}(\varphi(e_\nu \otimes e_{-\nu}) - \varepsilon\varphi(e_{-\nu} \otimes e_\nu))$$

$$= \bar{1} \sum_{\substack{s>\nu \\ s\equiv\nu\,(\mathrm{mod}\ 2\ell)}} c_s \; - \bar{1} \sum_{\substack{s>0 \\ s\equiv-\nu+1\,(\mathrm{mod}\ 2\ell)}} c_s + \bar{1}c_\nu - \frac{k}{2}\,\bar{1}$$

so

$$\Gamma(s_{e_\nu,e_{-\nu}} \otimes t^0) = \Big(\sum_{\substack{s \geq \nu \\ s \equiv \nu (\mathrm{mod}\ 2\ell)}} c_s - \sum_{\substack{s>0 \\ s \equiv 1-\nu (\mathrm{mod}\ 2\ell)}} c_s - \frac{k}{2}\Big) \Gamma,$$

and

$$\Gamma d(0) = \sum_{\nu=1}^{\ell} \Gamma D_\nu(0)$$

$$= \sum_{\nu=1}^{\ell} \sum_{j>0} j\Gamma(\varphi(e_{\nu+2\ell j} \otimes e_{-\nu-2\ell j}) + \varphi(e_{2\ell j+1-\nu} \otimes e_{-2\ell j-1+\nu})) + \frac{k\ell}{8} \Gamma$$

$$= \Big(\Big(\sum_{\nu=1}^{\ell} \Big(\sum_{j>0} jc_{\nu+2j\ell} + jc_{2j\ell+1-\nu}\Big)\Big) + \frac{k\ell}{8}\Big)\Gamma ,$$

all the sums being finite. Thus Γ is a weight vector.

Next for $i > 0$,

$$(m\varphi(e_{-(i+1)} \otimes e_i))(S_{e_\nu,e_{-\nu}} \otimes t^0)$$

$$= \sum_{j>1} m[\varphi(e_{-(i+1)} \otimes e_i),\varphi(e_{\nu+2j\ell} \otimes e_{-\nu-2j\ell}) - \varphi(e_{2\ell j+1-\nu} \otimes e_{\nu-2\ell j-1})]$$

$$+ \frac{1}{2} m[\varphi(e_{-(i+1)} \otimes e_i),\varphi(e_\nu \otimes e_{-\nu}) - \varepsilon\varphi(e_{-\nu} \otimes e_\nu)] + m(S_{e_\nu,e_{-\nu}} \otimes t^0)\varphi(e_{-i-1} \otimes e_i)$$

and the sum of commutators is equal to 0 unless $i \equiv \nu,\ 1-\nu,\ \nu-1,\ -\nu \bmod 2j\ell$. In these cases it is $-m\varphi(e_{-i-1} \otimes e_i)$, $m\varphi(e_{-i-1} \otimes e_i)$, $-m\varphi(e_{-i-1} \otimes e_i)$, respectively. Likewise,

$$m[\varphi(e_{-i-1} \otimes e_i),d(0)] = \sum_{\nu=1}^{\ell} \sum_{j>0} jm[\varphi(e_{-i-1} \otimes e_i),\varphi(e_{2\ell j+\nu} \otimes e_{-\nu-2\ell j}) + \varphi(e_{2\ell j+1-\nu} \otimes e_{-2\ell j-1+\nu})].$$

If $i \not\equiv \ell\ (\mathrm{mod}\ 2\ell)$, one sees that this is zero, while for $i \equiv \ell\ (\mathrm{mod}\ 2\ell)$ it is

$$m\varphi(e_{-i-1} \otimes e_i) .$$

It follows that if m is an H–weight vector, so is $m\varphi(e_{-i-1} \otimes e_i)$ for each $i > 0$. If $\varepsilon = -1$,

$$m[\varphi(e_{-1} \otimes e_{-1}), S_{e_\nu, e_{-\nu}} \otimes t^0] = 0$$

unless $\nu = 1$, where the value is

$$\tfrac{1}{2}m[\varphi(e_{-1} \otimes e_{-1}), \varphi(e_1 \otimes e_{-1}) + \varphi(e_{-1} \otimes e_1)] = 2m\varphi(e_{-1} \otimes e_{-1}),$$

and $m[\varphi(e_{-1} \otimes e_{-1}), d(0)] = 0$, so that again $m\varphi(e_{-1} \otimes e_{-1})$ is a weight vector if m is. Finally, if $\varepsilon = 1$, $m[\varphi(e_{-1} \otimes e_{-2}), S_{e_\nu, e_{-\nu}} \otimes t^0] = 0$ unless $\varepsilon = 1$ or 2, in which cases the value is

$$\tfrac{1}{2}m[\varphi(e_{-1} \otimes e_{-2}), \varphi(e_\nu \otimes e_{-\nu}) - \varphi(e_{-\nu} \otimes e_\nu)] = m\varphi(e_{-1} \otimes e_{-2}),$$

and $m[\varphi(e_{-1} \otimes e_{-2}), d(0)] = 0$. If follows that M_Λ is spanned by H–weight vectors.

Recall from [4] that a K–module M is called integrable if it is spanned by H–weight vectors and if each $x_{\pm\beta_\nu}$, $0 \le \nu \le \ell$, acts locally nilpotently on M. We have seen above that the first of these conditions is satisfied by M_Λ. For $1 \le \nu < \ell$, x_{β_ν} is represented by

$$m \mapsto \sum_{j>0} m(\varphi(e_{-(\ell-\nu)}(j) \otimes e_{\ell-\nu+1}(-j)) - \varepsilon\varphi(e_{\ell-\nu+1}(j) \otimes e_{-(\ell-\nu)}(-j)))$$

$$+ \tfrac{1}{2}m(\varphi(e_{-(\ell-\nu)}(0) \otimes e_{\ell-\nu+1}(0)) - \varepsilon\varphi(e_{\ell-\nu+1}(0) \otimes e_{-(\ell-\nu)}(0))).$$

From the relations in A_k, all the operators $\varphi(e_{-(\ell-\nu)}(j) \otimes e_{\ell-\nu+1}(-j))$ $(j \ge 0)$ commute, as do all the $\varphi(e_{\ell-\nu+1}(j) \otimes e_{-(\ell-\nu)}(-j))$, $j \ge 0$, and those of the first type commute with those of the second. It follows that if at most r of these terms fail to annihilate m, then $mx_{\beta_\nu}^s = 0$ if $s > rk$.

A similar argument applies in all the other cases, so that M_Λ is an integrable K–module.

From the computations above, we see that we can be more precise in specifying the H–weight of $m\varphi(e_{-(i+1)} \otimes e_i)$, etc., when m is an H–weight vector of weight λ. Namely, by comparing values at our basis of H with those of the fundamental roots $\beta_0, \ldots, \beta_\ell$, we find

Lemma 2. Let m be an H–weight vector of M_Λ of weight λ. Then if $1 < \nu < \ell$, and if $i > 0$, $i \equiv \pm\nu \pmod{2\ell}$, $m\varphi(e_{-(i+1)} \otimes e_i)$ is an H–weight vector of weight $\lambda - \beta_{\ell-\nu}$.

(a) If $\varepsilon = -1$ and $i > 0$, $i \equiv 0 \pmod{2\ell}$, then $m\varphi(e_{-(i+1)} \otimes e_i)$ is an H–weight vector of weight

$\lambda - \beta_\ell$, as is $m\varphi(e_{-1} \otimes e_{-1})$.

(b) If $\varepsilon = -1$ and $i > 0, i \equiv \ell \pmod{2\ell}$, then $m\varphi(e_{-(i+1)} \otimes e_i)$ is an *H*–weight vector of weight $\lambda - \beta_0$.

(c) If $\varepsilon = 1$ and $i > 0, i \equiv 0 \pmod{2\ell}$, then $m\varphi(e_{-(i+1)} \otimes e_i)$ is an *H*–weight vector of weight $\lambda - \beta_{\ell-1} - \beta_\ell$, and $m\varphi(e_{-1} \otimes e_{-2})$ is an *H*–weight vector of weight $\lambda - \beta_\ell$.

(d) If $\varepsilon = 1$ and $i > 0, i \equiv \ell \pmod{2\ell}$, then $m\varphi(e_{-(i+1)} \otimes e_i)$ is an *H*–weight vector of weight $\lambda - \beta_0 + \beta_1$.

(In the statement of the lemma, "0" is understood to be a weight vector of every weight.)

Thus if λ is the *H*–weight of I, as determined in the calculations earlier in this section, it follows for $\varepsilon = -1$ that all *H*–weights of M_Λ have the form $\lambda - \sum_{i=0}^{\ell} n_i\beta_i$, where the n_i are non–negative integers. For $\varepsilon = 1$ we can only conclude that all *H*–weights of M_Λ have the form

$$\lambda - \sum_{i=1}^{\ell} n_i\beta_i - n_0(\beta_0 - \beta_1),$$

the n_i being as above. For this reason we impose on Λ an additional restriction when $\varepsilon = 1$, namely that $m_i = 0$ for all $i \equiv \ell \pmod{2\ell}$. Then we have

<u>Proposition 2</u>. Let k, M_Λ,... be as above. If $\varepsilon = 1$, assume moreover that $m_i = 0$ for all $i \equiv \ell \pmod{2\ell}$. Then M_Λ is a sum of *H*–weight spaces for weights of the form $\lambda - \sum_{i=0}^{\ell} n_i\beta_i$, where λ is the H–weight of the generator I of M_Λ (as $A_k(V_\infty)$non–negative integers. Each *H*–weight space of M_Λ is finite–dimensional. In terms of the fundamental *H*–weights $\omega_0,\dots,\omega_\ell$ associated with extended diagrams in the labeling of §1, the restriction of λ to the subspace H' of H generated by the "coroots" is equal to $\sum_{i=0}^{\ell} d_i\omega_i$, where

$$d_{\ell-\nu} = \sum_{i\equiv 1\pm\nu \pmod{2\ell}} m_i, \text{ for } 1 \le \nu < \ell,$$

and, for $\varepsilon = -1$,

$$d_\ell = \sum_{i\equiv 1 \pmod{2\ell}} m_i$$

$$d_0 = \sum_{i\equiv \ell+1 \pmod{2\ell}} m_i ;$$

for $\varepsilon = 1$,

$$d_\ell = m_1 + \sum_{\substack{i>2 \\ i\equiv 2 \pmod{2\ell}}} m_i + 2\sum_{\substack{i>1 \\ i\equiv 1 \pmod{2\ell}}} m_i ,$$

$$d_0 = 2\sum_{i\equiv \ell+1 \pmod{2\ell}} m_i + \sum_{i\equiv \ell+2 \pmod{2\ell}} m_i .$$

Proof. The first assertion, for $\varepsilon = -1$, follows at once from Lemma 2 and from the fact that M_Λ is linearly generated by elements of the form (1). To see it when $\varepsilon = 1$, subject to our restrictions, it is sufficient by Lemma 2 to prove by induction on t, that each non-zero element of M_Λ of the form (1) belongs to an H-weight of the asserted form. This is known for $t = 0$, so we may assume $t > 0$ is minimal for which it is not proved. If $i_t \not\equiv \ell \pmod{2\ell}$, the assertion follows by inductive hypothesis and Lemma 2. Thus we may assume $i_t \equiv \ell \pmod{2\ell}$, and further that $i_t, i_{t-1}, \ldots, i_{t-\nu}$ are congruent to ℓ, while $i_{t-\nu-1}$ is not; we must have $\nu < t-1$ by the assumptions. If $i_{t-\nu-1} \not\equiv \ell \pm 1 \bmod 2\ell$, or if $|i_{t-\nu-1} - i_s| > 1$ for all $s \geq t-\nu$, then $X'_{-\alpha_{i_{t-\nu-1}}}$ commutes with all succeeding $X'_{\alpha_{i_s}}$, and $\mathrm{I}X'_{-\alpha_{i_1}} \ldots X'_{-\alpha_{i_{t-\nu-2}}} X'_{-\alpha_{i_{t-\nu}}} \ldots X'_{-\alpha_{i_t}} \neq 0$ is a weight vector for a weight of the prescribed form, by inductive hypothesis. Our assertion then follows from Lemma 2. Accordingly, we may assume $|i_{t-\nu-1} - i_s| = 1$ for some $s \geq t-\nu$; by commuting (commutative) factors $X'_{-\alpha_i}$, we may assume $s = t-\nu$. Then $[X'_{-\alpha_{i_{t-\nu-1}}}, X'_{-\alpha_{i_{t-\nu}}}]$ commutes with all succeeding $X'_{-\alpha_{i_s}}$, so (1) is equal to

$$\mathrm{I}X'_{-\alpha_{i_1}} \ldots X'_{-\alpha_{i_{t-\nu-2}}} X'_{-\alpha_{i_{t-\nu+1}}} \ldots X'_{-\alpha_{i_t}} [X'_{-\alpha_{i_{t-\nu-1}}}, X'_{-\alpha_{i_{t-\nu}}}]$$

$$+ \mathrm{I}X'_{-\alpha_{i_1}} \ldots X'_{-\alpha_{i_{t-\nu-2}}} X'_{-\alpha_{i_{t-\nu}}} X'_{-\alpha_{i_{t-\nu-1}}} X'_{-\alpha_{i_{t-\nu+1}}} \ldots X'_{-\alpha_{i_t}} .$$

If the first term here is non-zero, it belongs to a weight of the asserted form by the inductive hypothesis and Lemma 2. If the second term is non-zero, an induction on ν (and Lemma 2) completes the argument.

To see that H–weight spaces are finite–dimensional, let $\mu = \lambda - \sum_{i=0}^{\ell} s_i\beta_i$ be an H–weight, where the s_i are non–negative integers, and let an element (1) be different from zero and belong to μ. By construction of M_Λ, there is an $r > 0$ such that $\mathrm{T}X'_{-\alpha_i} = 0$ for all $i > r$. If $\varepsilon = -1$, let $s = \sum_{i=i}^{\ell} s_i$; if $\varepsilon = 1$, set $s > 2(s_2+s_3+...+s_\ell) + 1$. We claim that if an element (1) is non–zero and belongs to the weight μ, then $i_j \le r+s$ for all j; the finite–dimensionality of the space of H–weight μ then follows from the local–finiteness of $A_k(V_\infty)$.

Thus suppose some $i_j > r+s$. This index cannot be i_1 because $i_j > r$, and we may assume it is the first of such indices ($> r+s$). Because $X'_{-\alpha_{i_j}}$ commutes with all $X'_{-\alpha_{i_\nu}}$, $|i_j - i_\nu| \neq 1$ (for $\varepsilon = 1$ this holds if $i_j > 2$, a situation we may safely assume), and annihilates T, we may assume an $X'_{-\alpha_{(i_j \pm 1)}}$ precedes $X'_{-\alpha_{i_j}}$ somewhere in (1). The index here is greater than $r+s-1$, and by our first assumption it is equal to i_j-1. Repeating, we may assume that in (1) all the indices $r+1, r+2, ..., i_j$ occur and in this order. If $\varepsilon = -1$ it follows by Lemma 2 that (1) belongs to an H–weight $\lambda - \sum_{i=0}^{\ell} u_i\beta_i$, where $\Sigma\, u_i \ge (i_j - r) > s$, therefore not to μ, by the linear independence of the β_i. If $\varepsilon = 1$, at least $(i_j - r - 1)\frac{\ell-1}{\ell+1}$ of these indices ν have the property that $X'_{-\alpha_\nu}$ transforms a vector of H–weight σ into one of H–weight $\sigma - \beta_i$, where $i \ge 2$, and $(i_j - r - s)\frac{\ell-1}{\ell+1} > \frac{(s-1)}{2} > s_2 + ... + s_\ell$ (recall $\ell \le 3$). It follows from Lemma 2 that (1) here belongs to an H–weight as for $\varepsilon = 1$, here with $\sum_{i\ge 2} u_i > \sum_{i \ge 2} s_i$, and again we have a contradiction. This proves the finite–dimensionality of H–weight spaces.

The final assertion results from the calculation of $\mathrm{T}\,[x_{\beta_i}, x_{-\beta_i}]$, $0 \le i \le \ell$, using that the fundamental coroots $h_i^{\check{}}$ are the $[x_{\beta_i}, x_{-\beta_i}]$, as in the early paragraphs of this section. (For more details of this calculation, see §3.)

In particular we see that in any canonical ordering of the weights in which the β_i are positive, every H–weight is less than or equal to λ, and the H–weight space for λ has dimension 1. So far as this goes, we may therefore refer to λ as the highest H–weight of M_Λ, without being able to assert that T generates the K–module M_Λ.

Furthermore, $\lambda = \sum_{i=0}^{\ell} d_i\omega_i$ satisfies $2\sum_{i=0}^{\ell} d_i = k$, if $\varepsilon = -1$, and

$$d_0 + d_1 + 2\sum_{i=2}^{\ell-2} d_i + d_{\ell-1} + d_\ell = k$$

if $\varepsilon = 1$. In the latter case the condition that $m_j = 0$ for all $j \equiv \ell \pmod{2\ell}$ has the effect of making $d_0 \geq d_1$.

§3. Hermitian Form and Complete Reducibility.

When $\mathbb{C}$ is the ground field, it has been established in [6] (§5) that there is a positive-definite hermitian form h on M_Λ, associated with a conjugate-linear mapping $f\colon V_\infty \to V_\infty$ such that for all $m,m' \in M_\Lambda$; all $u,v \in V_\infty$,

$$h(m\varphi(u \otimes v),m') = h(m,m'\varphi(f(v) \otimes f(u))) . \tag{4}$$

Comparison of the definition of f in the reference with the construction of our basis $\{e_i\}$ show that $f(e_\nu(j)) = e_{-\nu}(-j)$ whenever $1 \leq \nu \leq \ell$, while $f(e_{-\nu}(j)) = \varepsilon e_\nu(-j)$. It follows that if $m \in M_\Lambda$, $n \in \mathbb{Z}$, ξ and $\eta \in \mathbb{C}$, and μ,ν are indices from $\{\pm1,\ldots,\pm\ell\}$, then

$$\sum_{j>\frac{n}{2}} m\{(\varphi(f(\eta e_\nu(n-j)) \otimes f(\xi e_\mu(j))) - \varepsilon\varphi(f(\xi e_\mu(n-j)) \otimes f(\eta e_\nu(j)))\}$$

$$+ \tfrac{1}{2}m(\varphi(f(\eta e_\nu(\tfrac{n}{2})) \otimes f(\xi e_\mu(\tfrac{n}{2}))) - \varepsilon\varphi(f(\xi e_\mu(\tfrac{n}{2})) \otimes f(\eta e_\nu(\tfrac{n}{2}))))] =$$

$$\pm \sum_{j'>-\frac{n}{2}} \bar{\xi}\bar{\eta} m\{(\varphi(e_{-\nu}(j') \otimes e_{-\mu}(-n-j')) - \varepsilon\varphi(e_{-\mu}(j') \otimes e_{-\nu}(-n-j))\}$$

$$+ \tfrac{1}{2}\bar{\xi}\bar{\eta} m(\varphi(e_{-\nu}(\tfrac{-n}{2}) \otimes e_{-\mu}(\tfrac{-n}{2})) - \varepsilon\varphi)e_{-\mu}(\tfrac{-n}{2}) \otimes e_{-\nu}(\tfrac{-n}{2})))$$

makes sense, and is equal to

$$\pm\, m(S_{(\bar{\eta}e_\nu),(\bar{\xi}e_\mu)} \otimes t^{-n}) .$$

That is, we have a conjugate-linear map σ of the Lie algebra which is the image of K' in

End M_Λ, to itself, sending (the image of) $S_{u,v} \otimes t^n$ to (the image of) $\pm S_{f(v),\ f(u)} \otimes t^{-n}$. We let σ fix "c", or its image $-\frac{k}{2}I$, and from the properties of h we have for all $m,m' \in M_\Lambda$, $x \in \mathrm{Im}(K')$,

$$h(mx,m') = h(m,m'\sigma(x)) .$$

From this it follows that if N is a K'–submodule of M_Λ, so is the orthogonal space $N^\perp$. Similar considerations apply to the Virasoro elements, where we find

$$h(md(n),m') = h(m,m'd(-n))$$

for all n. Thus, in particular, if N is a K–submodule of M_Λ, so is $N^\perp$.

From the fact that the form is positive–definite, $N \cap N^\perp = \{0\}$. Furthermore, in the restriction of the form to H–weight spaces, which are finite–dimensional under the assumptions of §2, we see that each weight space W has $W = (W \cap N) \oplus (W \cap N^\perp)$, so that $M_\Lambda = N \oplus N^\perp$. That is, M_Λ is a completely reducible K–module, as well as being a completely reducible B–module.

Let M' be the K'–submodule of M_Λ generated by I. Then I is an eigenvector of d(0) and d(0) normalizes K'. thus M' is also a K–submodule. By the remarks of [4] (p. 121, 1st ed.) it follows that M' is the irreducible integrable K–module of highest weight λ (as given below), and that

$$M_\Lambda = M' \oplus (M')^\perp ,$$

the last by the above.

The "fundamental coroots" for our roots $\beta_0,\ldots,\beta_\ell$ are

$$h_i^\vee = (S_{e_i,e_{-i}} - S_{e_{i+1},e_{-i-1}}) \otimes t^0, \qquad 1 \le i < \ell;$$

for $\varepsilon = -1$:

$$h_\ell^\vee = -S_{e_1,e_{-1}} \otimes t^0,$$

$$h_0^\vee = (S_{e_\ell,e_{-\ell}} \otimes t^0) - c ,$$

with $2 \sum_{i=0}^{\ell} h_i^\vee = -2c$, represented by $k\,\mathrm{I}$ on M_Λ.

For $\varepsilon = 1$,

$$h_\ell^\vee = -(S_{e_1,e_{-1}} + S_{e_2,e_{-2}}) \otimes t^0, \qquad h_0^\vee = (S_{e_{\ell-1},e_{-(\ell-1)}} + S_{e_\ell,e_{-\ell}}) \otimes t^0 - 2c,$$

with $h_0^{\check{}} + h_1^{\check{}} + 2\sum_{i=2}^{\ell-2} h_i^{\check{}} + h_{\ell-1} + h_\ell = -2c$, represented by $k\mathrm{I}$.

The highest weight λ of the K–module M' has values at these $h_i^{\check{}}$ that can be computed from the above. They are, for $\varepsilon = -1$,

(5) $$\lambda(h_i^{\check{}}) = \sum_{t\equiv\ell+1\pm i(\mathrm{mod}\ 2\ell)} m_t \quad , \quad 1 \le i < \ell,$$

$$\lambda(h_\ell^{\check{}}) = \sum_{t\equiv 1(\mathrm{mod}\ 2\ell)} m_t \quad ; \quad \lambda(h_0^{\check{}}) = \sum_{t\equiv\ell+1(\mathrm{mod}\ 2\ell)} m_t \quad ,$$

which of necessity satisfy $2\sum_{i=0}^{\ell} \lambda(h_i^{\check{}}) = k$.

Conversely given non–negative integers $b_0,\dots,b_\ell$ such that $2\Sigma\, b_i = k$, it is clear that we can find $m_1, m_2, \dots$ as above with $2\sum_{i=1}^{\ell} m_i = k$ and satisfying the relations (5) with $\lambda(h_i^{\check{}}) = b_i$. Indeed infinitely many such choices are possible for specified $b_0,\dots,b_\ell$. That is, when $\varepsilon = -1$, every irreducible integrable K–module, whose highest weight has a value at $d(0)$ of the form given in §2, occurs as a module M' from our construction.

For $\varepsilon = 1$, we find as before

$$\lambda(h_i^{\check{}}) = \sum_{t\equiv\ell+1\pm i\ (\mathrm{mod}\ 2\ell)} m_t \quad . \quad 1 \ge i < \ell;$$

$$\lambda(h_\ell^{\check{}}) = m_1 + \sum_{\substack{t>2\\ t\equiv 2(\mathrm{mod}\ 2\ell)}} m_t + 2\sum_{\substack{t>1\\ t\equiv 1(\mathrm{mod}\ 2\ell)}} m_t + \sum_{t\equiv\ell(\mathrm{mod}\ 2\ell)} m_t \quad ;$$

$$\lambda(h_0^{\check{}}) = \sum_{t\equiv\ell(\mathrm{mod}\ 2\ell)} m_t + 2\sum_{t\equiv\ell+1(\mathrm{mod}\ 2\ell)} m_t + \sum_{t\equiv\ell+2(\mathrm{mod}\ 2\ell)} m_t \quad ,$$

with

$$\lambda(h_0^{\check{}}) + \lambda(h_1^{\check{}}) + 2\sum_{i\ 2}^{\ell-2} \lambda(h_i^{\check{}}) + \lambda(h_{\ell-1}^{\check{}}) + \lambda(h_\ell^{\check{}}) = m_1 + m_2 + 2\sum_{t\ge 3} m_t = k.$$

From the constraint $m_t = 0$ for all $t \equiv \ell(\mathrm{mod}\ 2\ell)$ that we have imposed, $\lambda(h_1^{\check{}}) =$

$\sum_{t=\ell+2} m_t = \lambda(h_0^\check{}) - 2 \sum_{t=\ell+1} m_t$. Thus the highest weight λ of M' must satisfy

$$\lambda(h_0^\check{}) \geq \lambda(h_1^\check{}), \qquad \lambda(h_0^\check{}) \equiv \lambda(h_1^\check{}) (\text{mod } 2)\,.$$

As for $\varepsilon = -1$, it is now an easy matter to see that if $b_0, b_1, \ldots, b_\ell$ are non-negative integers satisfying $b_0 + b_1 + 2\sum_{i=2}^{\ell-2} b_i + b_{\ell-1} + b_\ell = k$, $b_0 \equiv b_1 (\text{mod } 2)$, $b_0 \geq b_1$, then the irreducible integrable K'-module of highest weight λ with $\lambda(h_i^\check{}) = b_i$ for all i arises as the module M' of our construction.

The results of these considerations can be summarized as follows:

Theorem. Let V be a complex vector space of dimension 2ℓ, with a given non-degenerate symmetric or alternate bilinear form. We further assume $\ell \geq 3$ if the form is symmetric. Let $\mathfrak{g}$ be the Lie algebra of skew transformations of V, and let

$$K = \mathfrak{g} \otimes \mathbb{C}[t,t^{-1}] \oplus \mathbb{C}c \oplus \mathbb{C}d(0)$$

be the associated Kac–Moody algebra (of first kind), normalized as in §1. Let k be a non-negative integer, with k even if the form is alternate, and let $b_0, \ldots, b_\ell$ be a set of non-negative integers with $b_0 + b_1 + b_{\ell-1} + b_\ell + 2\sum_{i=1}^{\ell-2} b_i = k$ and $b_0 \geq b_1$, $b_0 \equiv b_1 (\text{mod } 2)$, if the form on V is symmetric; with

$$2\sum_{i=0}^{\ell} b_i = k\,,$$

if the form on V is alternate. Then there is a simple right $A_k(V_\infty)$-module M with canonical generator I, a positive-definite hermitian scalar product on M consistent with the involution σ of §1 in $A_k(V_\infty)$, and an action of K on M defined by the formulas of §1. In this action the hermitian form on M is contravariant with respect to a canonical involution in K and M is a completely reducible integrable K-module, all weights of which are less than or equal to that of I. The weight spaces are finite-dimensional, and the K-submodule generated by I is an irreducible integrable K-module with I as highest weight vector. The highest weight λ satisfies $\lambda(h_i^\check{}) = b_i$

for all i, with canonical labeling as in §2.

§4. Bases and comparison of characters.

Here we speciialize to $\varepsilon = -1$ and $k = 2$, with a non–trivial module M for $A_2(V \otimes \mathbb{C}[t,t^{-1}])$, where V had dimension 2ℓ. Thus we have a symplectic basis $\{u_i \mid i \in Z, i \neq 0\}$ as before for our V_∞, and M is generated by a vector $v \neq 0$ such that $v\varphi(u_i \otimes u_j) = 0$ whenever $i+j > 0$, and such that exactly one of the quantities

$$(1) \quad \begin{cases} v(\varphi(u_{-i} \otimes u_i) - \varphi(u_{-(i+1)} \otimes u_{i+1})), & i \geq 1; \\ -\frac{1}{2}v \cdot (\varphi(u_{-1} \otimes u_1) + \varphi(u_1 \otimes u_{-1})) \end{cases}$$

is equal to v, the others in (1) being equal to zero.

In case it is the last quantity of (1) which is equal to v, it follows from $\varphi(u_1 \otimes u_{-1}) = \varphi(u_{-1} \otimes u_1) + 2$ that $v\varphi(u_{-1} \otimes u_1) = -2v$, $v\varphi(u_1 \otimes u_{-1}) = 0$. Then from

$$\varphi(u_{-i} \otimes u_i) - \varphi(u_{-(i+1)} \otimes u_{i+1}) = \varphi(u_i \otimes u_{-i}) - \varphi(u_{i+1} \otimes u_{-(i+1)})$$

in A_2, it follows that $v\varphi(u_i \otimes u_{-i}) = 0$ for all $i > 0$.

Continuing with this case, with v as above, it follows that if <u>either</u> $i > 0$ or $j > 0$, then $v\varphi(u_i \otimes u_j) \in Fv$, where F is the ground field. For this it suffices to assume $i+j < 0$, so $i > 0$, $j < -i$ may be assumed. From finite–dimensional theory, $\varphi(u_j \otimes u_{-(j+1)})$ annihilates v, as does $\varphi(u_i \otimes u_{j+1})$, by induction on $|i+j|$. The assertion follows by forming commutators.

Now it follows that M is linearly generated by elements

$$(2) \qquad v\varphi(u_{i_1} \otimes u_{i_2})\varphi(u_{i_3} \otimes u_{i_4}) \dots \varphi(u_{i_{2k-1}} \otimes u_{i_{2k}}),$$

where all of $i_1,\dots,i_{2k}$ are <u>negative</u>. Using identities in A_2 that are consequences of the defining identities, we can further normalize the representation (2):

<u>M is linearly generated by elements</u>

$$(3) \quad v\varphi(u_{s_1} \otimes u_{s_1})\varphi(u_{s_2} \otimes u_{s_2}) \dots (u_{s_j} \otimes u_{s_j}) \cdot \varphi(u_{t_1} \otimes u_{t_2})\varphi(u_{t_3} \otimes u_{t_4}) \dots \varphi(u_{t_{2k-1}} \otimes u_{t_{2k}}),$$

where $s_1,\dots,s_j, t_1,\dots,t_{2k}$ are distinct negative indices, subject to the conditions

(4) $$s_1 > s_2 > \dots > s_j; \quad t_1 > t_3 > \dots > t_{2k-1};$$
$$t_2 > t_4 > \dots > t_{2k}; \quad t_{2i-1} > t_{2i}, \text{ all } i.$$

Proposition. The elements (3), subject to the conditions (4) on the indices, are a basis for M.

The proof that I have is based on the observation that if m is an even positive integer, and if $b_k = \frac{1}{k+1}\binom{2k}{k}$ for each positive integer k, then the number of pairs of sequences of negative integers satisfying (4), with all entries $\geq -m$, is

(5) $$\sum_{i=0}^{\frac{m}{2}} \binom{m}{2i} 2^{m-2i} b_i = b_{m+1} = \frac{1}{m+2}\binom{2m+2}{m+1}.$$

[When a set S of integers of order $2i$ is fixed, the integer b_i is the number of sequences $t_1,\dots,t_{2i}$ satisfying the condition imposed in (4), and of length $2i$, that can be found from S.]

Now $b_{m+1} = \binom{2m}{m} - \binom{2m}{m-2}$ is the dimension of the irreducible module for the finite-dimensional symplectic Lie algebra corresponding to the subspace of V_∞ with basis $u_{-m},\dots,u_m$, and with highest weight that of V (for this Lie subalgebra $sp(V_m)$ of A_2. The expressions (3), subject to our constraints (4), with all indices $\geq -m$, are thus a basis for this $sp(V_m)$-module, so in particular are linearly independent.

The proof of the identity (5) involves counting certain classes of paths in one-dimensional random walks; the connection with the walks was pointed out to me by S. Kakutani.

I do not have a complete proof as yet that the set of linear generators given below for M in each of the other cases is linearly independent, but numerical evidence is quite clear. That is, we now let $n \geq 1$ be fixed, and suppose our generator v for the A_2-module M is fixed by $\varphi(u_{-n} \otimes u_n) - \varphi(u_{-(n+1)} \otimes u_{n+1})$, and annihilated by all the other elements (1). As above, we can prove that M is linearly generated by all elements

(6) $$v\varphi(u_{s_1} \otimes u_{s_1}) \dots \varphi(u_{s_j} \otimes u_{s_j})\varphi(u_{t_1} \otimes u_{t_2}) \dots \varphi(u_{t_{2k-1}} \otimes u_{t_{2k}}),$$

where $s_1,\dots,s_j, t_1,\dots,t_{2k}$ are distinct,

(7) $$s_j < s_{j-1} < \dots < s_1 < -n; \quad t_{2k-1} < t_{2k-3} < \dots < t_3 < t_1 \le n;$$

$$t_{2k} < t_{2k-2} < \dots < t_2 < -n; \quad t_{2i-1} > t_{2i}, \text{ all } i.$$

It seems likely that closer analysis of weight spaces in the finite–dimensional case will yield the linear independence of the elements (6), subject to the constraints (7).

One can attempt to study how close M is to being an irreducible K–module by comparing the K–character of M with that of the irreducible K–module of highest weight of the weight of the generator v. We illustrate this in the case where our claim as to the basis for M is satisfactorily proved, a case where the highest weight for K is the fundamental weight "ω_0". Without getting into particulars, let me simply say that comparison of the principally specialized characters of the irreducible K–module of highest weight ω_0 and the K–module M comes down to comparison of the following formal power series in q:

(1) $$\left(\prod_{j=1}^{\infty}(1-q^{2j-1})\right)^{-1} \cdot \prod_{\substack{j=1 \\ j\not\equiv\pm1,0(\mathrm{mod}\ \ell+2)}}^{\infty}(1-q^{2j})^{-1} = \sum_{n=0}^{\infty} p_\ell^*(n)q^n,$$

where $p_\ell^*(n)$ is the number of partitions of n into integers $\not\equiv 0,\pm2(\mathrm{mod}\ 2\ell+4)$; and

(2) $$\sum_{n=0}^{\infty} a_n q^n, \quad \text{where}$$

$$a_n = \sum_{j=0}^{[\frac{n}{2}]} \Sigma'_{p_j^{(1)}} \Sigma''_{p_{n-2j}^{(2)}} \frac{1}{m+1}\binom{2m}{m},$$

obtained as follows:

$\Sigma'_{p_j^{(1)}}$ sums over all partitions of j into a set of <u>distinct positive</u> integers; for each such partition, $\Sigma'_{p_{n-2j}^{(2)}}$ sums over all partitions of n–2j into a set of <u>distinct positive integers, disjoint from that of the partition</u> $p_j^{(1)}$, <u>and of even cardinality</u> 2m.

It will be noted that "ℓ" does not appear in 2), suggesting that the connection with the irreducible K–module is less close than one might hope for. For low values of ℓ, one checks that for some relatively small n, $a_n > p_\ell^*(n)$, in which cases our *K*–module *M* is definitely reducible.

One can reasonably expect to obtain the decomposition of *M* into irreducible *K*–submodules by writing the principally specialized character of *M* as a sum of principally specialized characters of irreducible *K*–modules. In case $\ell = 1$, preliminary calculations suggest that the highest weights include $\omega_0, \omega_0-2\delta, \omega_0-4\delta, \ldots$, where δ is the basic positive "null root". Perhaps each of these modules occurs with multiplicity one, but this seems more doubtful.

References

[1] Feingold, A. and Frenkel, I., Classical affine algebras. Advances in Math., 58 (1985), 117–172.

[2] Frenkel, I., Spinor representations of affine Lie algebras. Proc. N.A.S. (USA) 77 (1980), 6303–6306.

[3] ________, Two constructions of affine Lie algebra representations and boson–fermion correspondence in quantum field theory. J. Funct. Analysis 44 (1981), 259–327.

[4] Kac, V., Infinite Dimensional Lie Algebras, Birkhauser, Boston–Basel–Stuttgart, 1983. (2nd ed., Cambridge Univ. Press, 1985.)

[5] Kac, V. and Peterson, D., Spin and wedge representations of infinite–dimensional Lie algebras and groups. Proc. N.A.S. (USA) 78 (1981), 3308–3312.

[6] Seligman, G., Spin–like modules for certain infinite–dimensional Lie algebras. Séminaire d'algèbre P. Dubreil et M–P. Malliavin, 1985. Springer Lecture Notes in Math. 1220 (1986), pp. 33–55.

1980 AMS subject classification (1985 revision) 17B67, 17B65

This paper is in final form, and no version of it will be submitted elsewhere.

Workshop Lectures

Henning Andersen	Representations of Frobenius kernels of algebraic groups.
Georgia Benkart	Stability results for modules of sl(n), o(n), sp(n) as $n \longrightarrow \infty$.
Richard Block	On the classification of restricted Lie algebras. and On the Hopf algebra of representative functions of Lie algebras.
Alberto Elduque	On Lie algebras with subalgebras of codimension one.
Rolf Farnsteiner	Cohomology groups of Kac-Moody algebras.
James Humphreys	Tensor products of simple modules of characteristic p.
J. Marshall Osborn	On the simple Lie algebras of prime characteristic.
George Seligman	Representations of the Kac-Moody algebra $C_n^{(1)}$ and the Virasoro algebra.
Shirlei Serconek	Forms of restricted Lie algebras of Cartan type.
Peter Slodowy	A parabolic partition of Kac-Moody groups.
Helmut Strade	Some contributions to the classification problem.
Vicente Varea	On the subalgebra lattice of Lie algebras over perfect fields of prime characteristic.
Robert Wilson	On the recognition theorem I and II.

Workshop Participants

Henning Andersen	Aarhus Universitet
Georgia Benkart	University of Wisconsin
Stephen Berman	University of Saskatchewan
Richard Block	University of California, Riverside
Kevin Bowman	University of Lancaster
Daniel Britten	University of Windsor
Gordon Brown	University of Colorado
Keith Chavey	University of Wisconsin
Alberto Elduque	Universidad de Zaragoza
Rolf Farnsteiner	University of Wisconsin, Milwaukee
Thomas Gregory	Ohio State University, Mansfield
Mark Hall	Northeast Missouri State University
James Humphreys	University of Massachusetts
Ian Musson	University of Wisconsin, Milwaukee
Timo Neuvonen	University of Turku
J. Marshall Osborn	University of Wisconsin
Karl Peters	University of Wisconsin
John Schue	Macalester College
George Seligman	Yale University
Shirlei Serconek	Rutgers University
Peter Slodowy	University of Liverpool
Louis Solomon	University of Wisconsin
Helmut Strade	Universität Hamburg
Jeffrey Stroomer	University of Wisconsin
Vicente Varea	Universidad de Zaragoza
Qing Wang	University of Wisconsin
Robert Wilson	Rutgers University

LECTURE NOTES IN MATHEMATICS
Edited by A. Dold and B. Eckmann

Some general remarks on the publication of proceedings of congresses and symposia

ecture Notes aim to report new developments - quickly, informally nd at a high level. The following describes criteria and procedures hich apply to proceedings volumes. The editors of a volume are trongly advised to inform contributors about these points at an arly stage.

1. One (or more) expert participant(s) of the meeting should act as the responsible editor(s) of the proceedings. They select the papers which are suitable (cf. §§ 2, 3) for inclusion in the proceedings, and have them individually refereed (as for a journal). It should not be assumed that the published proceedings must reflect conference events faithfully and in their entirety. Contributions to the meeting which are not included in the proceedings can be listed by title. The series editors will normally not interfere with the editing of a particular proceedings volume - except in fairly obvious cases, or on technical matters, such as described in §§ 2, 3. The names of the responsible editors appear on the title page of the volume.

2. The proceedings should be reasonably homogeneous (concerned with a limited area). For instance, the proceedings of a congress on "Analysis" or "Mathematics in Wonderland" would normally not be sufficiently homogeneous.

   One or two longer survey articles on recent developments in the field are often very useful additions to such proceedings - even if they do not correspond to actual lectures at the congress. An extensive introduction on the subject of the congress would be desirable.

3. The contributions should be of a high mathematical standard and of current interest. Research articles should present new material and not duplicate other papers already published or due to be published. They should contain sufficient information and motivation and they should present proofs, or at least outlines of such, in sufficient detail to enable an expert to complete them. Thus resumes and mere announcements of papers appearing elsewhere cannot be included, although more detailed versions of a contribution may well be published in other places later.

   Surveys, if included, should cover a sufficiently broad topic, and should in general not simply review the author's own recent research. In the case of surveys, exceptionally, proofs of results may not be necessary.

   "Mathematical Reviews" and "Zentralblatt für Mathematik" require that papers in proceedings volumes carry an explicit statement that they are in final form and that no similar paper has been or is being submitted elsewhere, if these papers are to be considered for a review. Normally, papers that satisfy the criteria of the Lecture Notes in Mathematics series also satisfy this

.../...

requirement, but we would strongly recommend that the contributing authors be asked to give this guarantee explicitly at the beginning or end of their paper. There will occasionally be cases where this does not apply but where, for special reasons, the paper is still acceptable for LNM.

§4. Proceedings should appear soon after the meeeting. The publisher should, therefore, receive the complete manuscript within nine months of the date of the meeting at the latest.

§5. Plans or proposals for proceedings volumes should be sent to one of the editors of the series or to Springer-Verlag Heidelberg. They should give sufficient information on the conference or symposium, and on the proposed proceedings. In particular, they should contain a list of the expected contributions with their prospective length. Abstracts or early versions (drafts) of some of the contributions are very helpful.

§6. Lecture Notes are printed by photo-offset from camera-ready typed copy provided by the editors. For this purpose Springer-Verlag provides editors with technical instructions for the preparation of manuscripts and these should be distributed to all contributing authors. Springer-Verlag can also, on request, supply stationery on which the prescribed typing area is outlined. Some homogeneity in the presentation of the contributions is desirable.

Careful preparation of manuscripts will help keep production time short and ensure a satisfactory appearance of the finished book. The actual production of a Lecture Notes volume normally takes 6 -8 weeks.

Manuscripts should be at least 100 pages long. The final version should include a table of contents and as far as applicable a subject index.

§7. Editors receive a total of 50 free copies of their volume for distribution to the contributing authors, but no royalties. (Unfortunately, no reprints of individual contributions can be supplied.) They are entitled to purchase further copies of their book for their personal use at a discount of 33.3 %, other Springer mathematics books at a discount of 20 % directly from Springer-Verlag. Contributing authors may purchase the volume in which their article appears at a discount of 33.3 %.

Commitment to publish is made by letter of intent rather than by signing a formal contract. Springer-Verlag secures the copyright for each volume.

LECTURE NOTES

Springer
Springer-Verlag
Berlin Heidelberg New York
London Paris Tokyo Hong Kong

ESSENTIALS FOR THE PREPARATION
OF CAMERA-READY MANUSCRIPTS

e preparation of manuscripts which are to be reproduced by photo-fset require special care. Manuscripts which are submitted in tech-cally unsuitable form will be returned to the author for retyping. ere is normally no possibility of carrying out further corrections ter a manuscript is given to production. Hence it is crucial that e following instructions be adhered to closely. If in doubt, please nd us 1 - 2 sample pages for examination.

neral. The characters must be uniformly black both within a single aracter and down the page. Original manuscripts are required: pho-copies are acceptable only if they are sharp and without smudges.

request, Springer-Verlag will supply special paper with the text ea outlined. The standard TEXT AREA (OUTPUT SIZE if you are using a point font) is 18 x 26.5 cm (7.5 x 11 inches). This will be scale-duced to 75% in the printing process. If you are using computer pesetting, please see also the following page.

ke sure the TEXT AREA IS COMPLETELY FILLED. Set the margins so that ey precisely match the outline and type right from the top to the ttom line. (Note that the page number will lie outside this area). nes of text should not end more than three spaces inside or outside e right margin (see example on page 4).

pe on one side of the paper only.

acing and Headings (Monographs). Use ONE-AND-A-HALF line spacing in e text. Please leave sufficient space for the title to stand out early and do NOT use a new page for the beginning of subdivisons of apters. Leave THREE LINES blank above and TWO below headings of ch subdivisions.

acing and Headings (Proceedings). Use ONE-AND-A-HALF line spacing the text. Do not use a new page for the beginning of subdivisons a single paper. Leave THREE LINES blank above and TWO below hea-ngs of such subdivisions. Make sure headings of equal importance in the same form.

e first page of each contribution should be prepared in the same . The title should stand out clearly. We therefore recommend that editor prepare a sample page and pass it on to the authors ether with these instructions. Please take the following as an mple. Begin heading 2 cm below upper edge of text area.

MATHEMATICAL STRUCTURE IN QUANTUM FIELD THEORY

John E. Robert
Mathematisches Institut, Universität Heidelberg
Im Neuenheimer Feld 288, D-6900 Heidelberg

ase leave THREE LINES blank below heading and address of the hor, then continue with the actual text on the same page.

tnotes. These should preferable be avoided. If necessary, type m in SINGLE LINE SPACING to finish exactly on the outline, and se-ate them from the preceding main text by a line.

Symbols. Anything which cannot be typed may be entered by hand : BLACK AND ONLY BLACK ink. (A fine-tipped rapidograph is suitable fo this purpose; a good black ball-point will do, but a pencil wi not). Do not draw straight lines by hand without a ruler (not even : fractions).

Literature References. These should be placed at the end of each pa per or chapter, or at the end of the work, as desired. Type them wi single line spacing and start each reference on a new line. Foll "Zentralblatt für Mathematik"/"Mathematical Reviews" for abbreviat titles of mathematical journals and "Bibliographic Guide for Edito and Authors (BGEA)" for chemical, biological, and physics journal. Please ensure that all references are COMPLETE and ACCURATE.

IMPORTANT

Pagination. For typescript, number pages in the upper right-hand co ner in LIGHT BLUE OR GREEN PENCIL ONLY. The printers will insert t final page numbers. For computer type, you may insert page numbe (1 cm above outer edge of text area).

It is safer to number pages AFTER the text has been typed and corre ted. Page 1 (Arabic) should be THE FIRST PAGE OF THE ACTUAL TEXT. T Roman pagination (table of contents, preface, abstract, acknowledg ments, brief introductions, etc.) will be done by Springer-Verlag.

If including running heads, these should be aligned with the insi edge of the text area while the page number is aligned with the ou side edge noting that right-hand pages are odd-numbered. Runni heads and page numbers appear on the same line. Normally, the runni head on the left-hand page is the chapter heading and that on t right-hand page is the section heading. Running heads should not included in proceedings contributions unless this is being done co sistently by all authors.

Corrections. When corrections have to be made, cut the new text fit and paste it over the old. White correction fluid may also used.

Never make corrections or insertions in the text by hand.

If the typescript has to be marked for any reason, e.g. for provisi nal page numbers or to mark corrections for the typist, this can done VERY FAINTLY with BLUE or GREEN PENCIL but NO OTHER COLOR: the colors do not appear after reproduction.

COMPUTER-TYPESETTING. Further, to the above instructions, please no with respect to your printout that
- the characters should be sharp and sufficiently black;
- it is not strictly necessary to use Springer's special typi paper. Any white paper of reasonable quality is acceptable.

If you are using a significantly different font size, you shou modify the output size correspondingly, keeping length to bread ratio 1 : 0.68, so that scaling down to 10 point font size, yields text area of 13.5 x 20 cm (5 3/8 x 8 in), e.g.

Differential equations.: use output size 13.5 x 20 cm.

Differential equations.: use output size 16 x 23.5 cm.

Differential equations.: use output size 18 x 26.5 cm.

Interline spacing: 5.5 mm base-to-base for 14 point characters (sta dard format of 18 x 26.5 cm).
If in any doubt, please send us 1 - 2 sample pages for examinatio We will be glad to give advice.

. 1201: Curvature and Topology of Riemannian Manifolds. Pro-
dings, 1985. Edited by K. Shiohama, T. Sakai and T. Sunada. VII,
pages. 1986.

. 1202: A. Dür, Möbius Functions, Incidence Algebras and Power
ies Representations. XI, 134 pages. 1986.

. 1203: Stochastic Processes and Their Applications. Proceedings,
5. Edited by K. Itô and T. Hida. VI, 222 pages. 1986.

. 1204: Séminaire de Probabilités XX, 1984/85. Proceedings. Edité
J. Azéma et M. Yor. V, 639 pages. 1986.

. 1205: B.Z. Moroz, Analytic Arithmetic in Algebraic Number Fields.
177 pages. 1986.

. 1206: Probability and Analysis, Varenna (Como) 1985. Seminar.
ed by G. Letta and M. Pratelli. VIII, 280 pages. 1986.

. 1207: P.H. Bérard, Spectral Geometry: Direct and Inverse
blems. With an Appendix by G. Besson. XIII, 272 pages. 1986.

. 1208: S. Kaijser, J.W. Pelletier, Interpolation Functors and
lity. IV, 167 pages. 1986.

. 1209: Differential Geometry, Peñíscola 1985. Proceedings.
ed by A.M. Naveira, A. Ferrández and F. Mascaró. VIII, 306 pages.
6.

. 1210: Probability Measures on Groups VIII. Proceedings, 1985.
ed by H. Heyer. X, 386 pages. 1986.

. 1211: M.B. Sevryuk, Reversible Systems. V, 319 pages. 1986.

. 1212: Stochastic Spatial Processes. Proceedings, 1984. Edited
P. Tautu. VIII, 311 pages. 1986.

. 1213: L.G. Lewis, Jr., J.P. May, M. Steinberger, Equivariant
ble Homotopy Theory. IX, 538 pages. 1986.

. 1214: Global Analysis – Studies and Applications II. Edited by
G. Borisovich and Yu.E. Gliklikh. V, 275 pages. 1986.

. 1215: Lectures in Probability and Statistics. Edited by G. del Pino
R. Rebolledo. V, 491 pages. 1986.

. 1216: J. Kogan, Bifurcation of Extremals in Optimal Control. VIII,
pages. 1986.

. 1217: Transformation Groups. Proceedings, 1985. Edited by S.
kowski and K. Pawalowski. X, 396 pages. 1986.

. 1218: Schrödinger Operators, Aarhus 1985. Seminar. Edited by
Balslev. V, 222 pages. 1986.

. 1219: R. Weissauer, Stabile Modulformen und Eisensteinreihen.
147 Seiten. 1986.

. 1220: Séminaire d'Algèbre Paul Dubreil et Marie-Paule Malliavin.
ceedings, 1985. Edité par M.-P. Malliavin. IV, 200 pages. 1986.

. 1221: Probability and Banach Spaces. Proceedings, 1985. Edited
J. Bastero and M. San Miguel. XI, 222 pages. 1986.

. 1222: A. Katok, J.-M. Strelcyn, with the collaboration of
Ledrappier and F. Przytycki, Invariant Manifolds, Entropy and
ards; Smooth Maps with Singularities. VIII, 283 pages. 1986.

. 1223: Differential Equations in Banach Spaces. Proceedings,
5. Edited by A. Favini and E. Obrecht. VIII, 299 pages. 1986.

. 1224: Nonlinear Diffusion Problems, Montecatini Terme 1985.
minar. Edited by A. Fasano and M. Primicerio. VIII, 188 pages.
6.

. 1225: Inverse Problems, Montecatini Terme 1986. Seminar.
ted by G. Talenti. VIII, 204 pages. 1986.

. 1226: A. Buium, Differential Function Fields and Moduli of
ebraic Varieties. IX, 146 pages. 1986.

. 1227: H. Helson, The Spectral Theorem. VI, 104 pages. 1986.

. 1228: Multigrid Methods II. Proceedings, 1985. Edited by
Hackbusch and U. Trottenberg. VI, 336 pages. 1986.

. 1229: O. Bratteli, Derivations, Dissipations and Group Actions on
-algebras. IV, 277 pages. 1986.

. 1230: Numerical Analysis. Proceedings, 1984. Edited by J.-P.
nnart. X, 234 pages. 1986.

. 1231: E.-U. Gekeler, Drinfeld Modular Curves. XIV, 107 pages.
6.

Vol. 1232: P.C. Schuur, Asymptotic Analysis of Soliton Problems. VIII, 180 pages. 1986.

Vol. 1233: Stability Problems for Stochastic Models. Proceedings, 1985. Edited by V.V. Kalashnikov, B. Penkov and V.M. Zolotarev. VI, 223 pages. 1986.

Vol. 1234: Combinatoire énumérative. Proceedings, 1985. Edité par G. Labelle et P. Leroux. XIV, 387 pages. 1986.

Vol. 1235: Séminaire de Théorie du Potentiel, Paris, No. 8. Directeurs: M. Brelot, G. Choquet et J. Deny. Rédacteurs: F. Hirsch et G. Mokobodzki. III, 209 pages. 1987.

Vol. 1236: Stochastic Partial Differential Equations and Applications. Proceedings, 1985. Edited by G. Da Prato and L. Tubaro. V, 257 pages. 1987.

Vol. 1237: Rational Approximation and its Applications in Mathematics and Physics. Proceedings, 1985. Edited by J. Gilewicz, M. Pindor and W. Siemaszko. XII, 350 pages. 1987.

Vol. 1238: M. Holz, K.-P. Podewski and K. Steffens, Injective Choice Functions. VI, 183 pages. 1987.

Vol. 1239: P. Vojta, Diophantine Approximations and Value Distribution Theory. X, 132 pages. 1987.

Vol. 1240: Number Theory, New York 1984–85. Seminar. Edited by D.V. Chudnovsky, G.V. Chudnovsky, H. Cohn and M.B. Nathanson. V, 324 pages. 1987.

Vol. 1241: L. Gårding, Singularities in Linear Wave Propagation. III, 125 pages. 1987.

Vol. 1242: Functional Analysis II, with Contributions by J. Hoffmann-Jørgensen et al. Edited by S. Kurepa, H. Kraljević and D. Butković. VII, 432 pages. 1987.

Vol. 1243: Non Commutative Harmonic Analysis and Lie Groups. Proceedings, 1985. Edited by J. Carmona, P. Delorme and M. Vergne. V, 309 pages. 1987.

Vol. 1244: W. Müller, Manifolds with Cusps of Rank One. XI, 158 pages. 1987.

Vol. 1245: S. Rallis, L-Functions and the Oscillator Representation. XVI, 239 pages. 1987.

Vol. 1246: Hodge Theory. Proceedings, 1985. Edited by E. Cattani, F. Guillén, A. Kaplan and F. Puerta. VII, 175 pages. 1987.

Vol. 1247: Séminaire de Probabilités XXI. Proceedings. Edité par J. Azéma, P.A. Meyer et M. Yor. IV, 579 pages. 1987.

Vol. 1248: Nonlinear Semigroups, Partial Differential Equations and Attractors. Proceedings, 1985. Edited by T.L. Gill and W.W. Zachary. IX, 185 pages. 1987.

Vol. 1249: I. van den Berg, Nonstandard Asymptotic Analysis. IX, 187 pages. 1987.

Vol. 1250: Stochastic Processes – Mathematics and Physics II. Proceedings 1985. Edited by S. Albeverio, Ph. Blanchard and L. Streit. VI, 359 pages. 1987.

Vol. 1251: Differential Geometric Methods in Mathematical Physics. Proceedings, 1985. Edited by P.L. García and A. Pérez-Rendón. VII, 300 pages. 1987.

Vol. 1252: T. Kaise, Représentations de Weil et GL_2 Algèbres de division et GL_n. VII, 203 pages. 1987.

Vol. 1253: J. Fischer, An Approach to the Selberg Trace Formula via the Selberg Zeta-Function. III, 184 pages. 1987.

Vol. 1254: S. Gelbart, I. Piatetski-Shapiro, S. Rallis. Explicit Constructions of Automorphic L-Functions. VI, 152 pages. 1987.

Vol. 1255: Differential Geometry and Differential Equations. Proceedings, 1985. Edited by C. Gu, M. Berger and R.L. Bryant. XII, 243 pages. 1987.

Vol. 1256: Pseudo-Differential Operators. Proceedings, 1986. Edited by H.O. Cordes, B. Gramsch and H. Widom. X, 479 pages. 1987.

Vol. 1257: X. Wang, On the C*-Algebras of Foliations in the Plane. V, 165 pages. 1987.

Vol. 1258: J. Weidmann, Spectral Theory of Ordinary Differential Operators. VI, 303 pages. 1987.

Vol. 1259: F. Cano Torres, Desingularization Strategies for Three-Dimensional Vector Fields. IX, 189 pages. 1987.

Vol. 1260: N.H. Pavel, Nonlinear Evolution Operators and Semigroups. VI, 285 pages. 1987.

Vol. 1261: H. Abels, Finite Presentability of S-Arithmetic Groups. Compact Presentability of Solvable Groups. VI, 178 pages. 1987.

Vol. 1262: E. Hlawka (Hrsg.), Zahlentheoretische Analysis II. Seminar, 1984–86. V, 158 Seiten. 1987.

Vol. 1263: V.L. Hansen (Ed.), Differential Geometry. Proceedings, 1985. XI, 288 pages. 1987.

Vol. 1264: Wu Wen-tsün, Rational Homotopy Type. VIII, 219 pages. 1987.

Vol. 1265: W. Van Assche, Asymptotics for Orthogonal Polynomials. VI, 201 pages. 1987.

Vol. 1266: F. Ghione, C. Peskine, E. Sernesi (Eds.), Space Curves. Proceedings, 1985. VI, 272 pages. 1987.

Vol. 1267: J. Lindenstrauss, V.D. Milman (Eds.), Geometrical Aspects of Functional Analysis. Seminar. VII, 212 pages. 1987.

Vol. 1268: S.G. Krantz (Ed.), Complex Analysis. Seminar, 1986. VII, 195 pages. 1987.

Vol. 1269: M. Shiota, Nash Manifolds. VI, 223 pages. 1987.

Vol. 1270: C. Carasso, P.-A. Raviart, D. Serre (Eds.), Nonlinear Hyperbolic Problems. Proceedings, 1986. XV, 341 pages. 1987.

Vol. 1271: A.M. Cohen, W.H. Hesselink, W.L.J. van der Kallen, J.R. Strooker (Eds.), Algebraic Groups Utrecht 1986. Proceedings. XII, 284 pages. 1987.

Vol. 1272: M.S. Livšic, L.L. Waksman, Commuting Nonselfadjoint Operators in Hilbert Space. III, 115 pages. 1987.

Vol. 1273: G.-M. Greuel, G. Trautmann (Eds.), Singularities, Representation of Algebras, and Vector Bundles. Proceedings, 1985. XIV, 383 pages. 1987.

Vol. 1274: N. C. Phillips, Equivariant K-Theory and Freeness of Group Actions on C*-Algebras. VIII, 371 pages. 1987.

Vol. 1275: C.A. Berenstein (Ed.), Complex Analysis I. Proceedings, 1985–86. XV, 331 pages. 1987.

Vol. 1276: C.A. Berenstein (Ed.), Complex Analysis II. Proceedings, 1985–86. IX, 320 pages. 1987.

Vol. 1277: C.A. Berenstein (Ed.), Complex Analysis III. Proceedings, 1985–86. X, 350 pages. 1987.

Vol. 1278: S.S. Koh (Ed.), Invariant Theory. Proceedings, 1985. V, 102 pages. 1987.

Vol. 1279: D. Ieşan, Saint-Venant's Problem. VIII, 162 Seiten. 1987.

Vol. 1280: E. Neher, Jordan Triple Systems by the Grid Approach. XII, 193 pages. 1987.

Vol. 1281: O.H. Kegel, F. Menegazzo, G. Zacher (Eds.), Group Theory. Proceedings, 1986. VII, 179 pages. 1987.

Vol. 1282: D.E. Handelman, Positive Polynomials, Convex Integral Polytopes, and a Random Walk Problem. XI, 136 pages. 1987.

Vol. 1283: S. Mardešić, J. Segal (Eds.), Geometric Topology and Shape Theory. Proceedings, 1986. V, 261 pages. 1987.

Vol. 1284: B.H. Matzat, Konstruktive Galoistheorie. X, 286 pages. 1987.

Vol. 1285: I.W. Knowles, Y. Saitō (Eds.), Differential Equations and Mathematical Physics. Proceedings, 1986. XVI, 499 pages. 1987.

Vol. 1286: H.R. Miller, D.C. Ravenel (Eds.), Algebraic Topology. Proceedings, 1986. VII, 341 pages. 1987.

Vol. 1287: E.B. Saff (Ed.), Approximation Theory, Tampa. Proceedings, 1985–1986. V, 228 pages. 1987.

Vol. 1288: Yu. L. Rodin, Generalized Analytic Functions on Riemann Surfaces. V, 128 pages, 1987.

Vol. 1289: Yu. I. Manin (Ed.), K-Theory, Arithmetic and Geometry. Seminar, 1984–1986. V, 399 pages. 1987.

Vol. 1290: G. Wüstholz (Ed.), Diophantine Approximation and Transcendence Theory. Seminar, 1985. V, 243 pages. 1987.

Vol. 1291: C. Mœglin, M.-F. Vignéras, J.-L. Waldspurger, Correspondances de Howe sur un Corps p-adique. VII, 163 pages. 1987

Vol. 1292: J.T. Baldwin (Ed.), Classification Theory. Proceedings, 1985. VI, 500 pages. 1987.

Vol. 1293: W. Ebeling, The Monodromy Groups of Isolated Singularities of Complete Intersections. XIV, 153 pages. 1987.

Vol. 1294: M. Queffélec, Substitution Dynamical Systems – Spectra Analysis. XIII, 240 pages. 1987.

Vol. 1295: P. Lelong, P. Dolbeault, H. Skoda (Réd.), Séminaire d'Analyse P. Lelong – P. Dolbeault – H. Skoda. Seminar, 1985/1986 VII, 283 pages. 1987.

Vol. 1296: M.-P. Malliavin (Ed.), Séminaire d'Algèbre Paul Dubreil e Marie-Paule Malliavin. Proceedings, 1986. IV, 324 pages. 1987.

Vol. 1297: Zhu Y.-l., Guo B.-y. (Eds.), Numerical Methods for Partia Differential Equations. Proceedings. XI, 244 pages. 1987.

Vol. 1298: J. Aguadé, R. Kane (Eds.), Algebraic Topology, Barcelon 1986. Proceedings. X, 255 pages. 1987.

Vol. 1299: S. Watanabe, Yu.V. Prokhorov (Eds.), Probability Theor and Mathematical Statistics. Proceedings, 1986. VIII, 589 pages 1988.

Vol. 1300: G.B. Seligman, Constructions of Lie Algebras and the Modules. VI, 190 pages. 1988.

Vol. 1301: N. Schappacher, Periods of Hecke Characters. XV, 16 pages. 1988.

Vol. 1302: M. Cwikel, J. Peetre, Y. Sagher, H. Wallin (Eds.), Functio Spaces and Applications. Proceedings, 1986. VI, 445 pages. 1988.

Vol. 1303: L. Accardi, W. von Waldenfels (Eds.), Quantum Probabili and Applications III. Proceedings, 1987. VI, 373 pages. 1988.

Vol. 1304: F.Q. Gouvêa, Arithmetic of p-adic Modular Forms. VIII, 12 pages. 1988.

Vol. 1305: D.S. Lubinsky, E.B. Saff, Strong Asymptotics for Extrem Polynomials Associated with Weights on ℝ. VII, 153 pages. 1988.

Vol. 1306: S.S. Chern (Ed.), Partial Differential Equations. Procee ings, 1986. VI, 294 pages. 1988.

Vol. 1307: T. Murai, A Real Variable Method for the Cauchy Transform and Analytic Capacity. VIII, 133 pages. 1988.

Vol. 1308: P. Imkeller, Two-Parameter Martingales and Their Quadr tic Variation. IV, 177 pages. 1988.

Vol. 1309: B. Fiedler, Global Bifurcation of Periodic Solutions wi Symmetry. VIII, 144 pages. 1988.

Vol. 1310: O.A. Laudal, G. Pfister, Local Moduli and Singularities. 117 pages. 1988.

Vol. 1311: A. Holme, R. Speiser (Eds.), Algebraic Geometry, Su dance 1986. Proceedings. VI, 320 pages. 1988.

Vol. 1312: N.A. Shirokov, Analytic Functions Smooth up to t Boundary. III, 213 pages. 1988.

Vol. 1313: F. Colonius, Optimal Periodic Control. VI, 177 page 1988.

Vol. 1314: A. Futaki, Kähler-Einstein Metrics and Integral Invariants. 140 pages. 1988.

Vol. 1315: R.A. McCoy, I. Ntantu, Topological Properties of Spaces Continuous Functions. IV, 124 pages. 1988.

Vol. 1316: H. Korezlioglu, A.S. Ustunel (Eds.), Stochastic Analysis a Related Topics. Proceedings, 1986. V, 371 pages. 1988.

Vol. 1317: J. Lindenstrauss, V.D. Milman (Eds.), Geometric Aspects Functional Analysis. Seminar, 1986–87. VII, 289 pages. 1988.

Vol. 1318: Y. Felix (Ed.), Algebraic Topology – Rational Homoto Proceedings, 1986. VIII, 245 pages. 1988

Vol. 1319: M. Vuorinen, Conformal Geometry and Quasireg Mappings. XIX, 209 pages. 1988.